The Theory of Everything Else

The Theory of Everything Else

L K O'Neal

Published by Winds, 2024.

While every precaution has been taken in the preparation of this book, the publisher assumes no responsibility for errors or omissions, or for damages resulting from the use of the information contained herein.

THE THEORY OF EVERYTHING ELSE

First edition. March 15, 2024.

Copyright © 2024 L K O'Neal.

ISBN: 979-8227500373

Written by L K O'Neal.

Also by L K O'Neal

Away Home
The Theory of Everything Else

Table of Contents

To ideas,

and dreams,

and uncompromised thought

and achievment

Could your thoughts and dreams determine everything?

Dreams
from one dream many
from many one

I t is a dream. Everyone has them.

    It is plain, ordinary, yet vivid with remarkable detail.

It is largely unremembered, except... what is peculiar.

There is an unceasing sense that the dream is important. A dread that discovery is lost, and a suffocating responsibility to remember it. There is panic, a fear that the responsibility cannot be met, that what is lost with the dream is the most important thought or idea that might ever be... but this is not possible.

In the dream, there is a bird's eye view above a bay on a lake in the Adirondack Mountains of New York State, an everyday place during the summer. There are myriad details of familiar boats and docks and shorefront homes. It is completely unmemorable.

Except that it seems to be real. The places exist and the details of them spot-on.

The details are startlingly specific, far greater than exists or is noticed in an awake life, as if shot with a lens of greater dimension than eyes are capable of.

    There is yet some awareness of the dream from the awake mind,
    a shroud of

remembrance. Why for this dream when not for others?

The bird's eye view floats out over the middle of the bay and then to the shoreline, to a point overlooking the shorefront homes and into backyards.

Here is unspectacular scenery of spaces not attended to, old damaged boats, piles of unused stone, brick and old timber lying about unorganized for any purpose.

Except for the vigorous detail.

Is this the point?

There are beer cups with ashes in them lying about a backyard deck, a rotting back porch with screening, a half finished patio, a dead bird crushed from flying into a floor to ceiling family room window.

The awake mind is gaining ground. In sleep it is another voice, an external one with marginal control, invading the circumstances of the dream, looking into it without understanding.

The circumstances begin to dissolve as the presence of the awake mind emerges. Will all of them?

There is understanding about the dream, broader than the dream, a recognition of something, a "knowing."

It is huge, gnawing, a mystery. Not just any mystery, THE mystery.

The dissolution is wrong. It invokes panic. The emotion incites awakeness. And so awakeness comes, the "knowing" still incomplete.

Uninvited, the awake mind intrudes further to resolve the mystery. But the more emotion, the more awake comes, the more the dream fades to black, the more the mystery dims in the spacelessness of sleep.

The dream dissipates in the dark as if a ghost. Perhaps this is where, after all, ghosts come from. It is like seeing something shiny deep underwater, something fantastic, discoverable, retrievable, but oxygen is limited, a return to the surface necessary. The object dims slowly, as if sinking, until it disappears with a foreboding that it is gone forever.

Even in this dream-state, mostly asleep, there is dread, the unbearable responsibility for the "knowing," for the not knowing.

As if the dream is everything. How could it be?

So awake must not come. An ordinary dream that is not must continue. The crushing weight, the importance of understanding must resolve, but nothing resolves in dreams.

The awakening comes with few memories...

and a howling loss of understanding.

But there is, yet, a flicker; knowledge of understanding but not the understanding itself.

Awake there is unknowledge; of sleep, of dreams, perhaps of so much more.

There is a time between asleep and awake. Time to think about the dream, not to dream it again. It is not that kind.

In this between-time, in dream shadows, awake is near.

An awake thought interludes as the dream vanishes.

There is the floating omniscient viewpoint, the fact of the great detail, but none of it, the witness of spaces and items intimately familiar and many never seen, but the specifics disappear.

Awakening cannot be stopped. It is unrelenting. Its very emergence, its eagerness, disassembles that which it wishes to understand. But this place between sleep and awake is a kind of Neverland for thought.

May it last is the plea. Let calm retrench and the mind drift back asleep.

It works a little.

Enough to remember something, a notion, an idea.

It comes slowly and all at once in that between time, a paradox.

The idea is simple, and gravely profound.

Everything in the dream exists.

Every exacting detail is accurate.

This is known, understood as fact, without the proof of it. Is this the "knowing?"

There is an astonishing conclusion; the dream displays known places and items accurately so it must also for those unknown just as accurately.

Yet, it is more.

Sometimes, there is no proof.

Sometimes there is just knowledge and no explanation. But can this be true?

In the awake mind's intervention, its small understanding touches a larger more massive body of knowledge, like reading a page of Wikipedia and being aware of an infinite amount more.

Awake, understanding withers to ordinary.

Importance is not evident.

If it is true that in this dream what had never been exposed or observed before is accurate...

The awake mind, ever logical, objects.

The dream is normal, not unlike dreams that everyone has all the time. There is nothing special about it. It is like all the others. It does not relate to the real world. Sound, simple awake logic.

But the dream's images are sharper, more crisp than eyes can record. The details of the setting more exact than real life and greater than any movie set could replicate. The greater dimension of the dream transposes real life from the awake world to the dream world.

There is no account for every single detail in the sweep of any given place in waking lives, but somehow there is in dreams.

The evidence of the dream disturbs the rational awake mind.

The dream seems to offer an omniscient viewpoint. The impression is made, therefore, that dreams can be "realer than real."

Are the details of dreams real or are they made-up? They are, of course, made up, at least many of them. Everyone understands this.

But even the awake mind recognizes that this is a significant question. If they are made-up, dreams are fantasy, fabricated for mysterious purposes possibly of biology such as re-energizing the brain, and perhaps rebooting given the circumstances of the day. This may be the purpose of sleep.

But, if sometimes they are real...?

The dream is "real," at least in part. It is set of real things, and real places that exist in the real world. But there is the fantasy of the omniscient viewpoint.

Prevailing sentiment labels dreams as fantasy, a mere biological function of sleep, perhaps to nourish and re-charge the brain. Science does not know the purpose of dreams. Any life-like or real depiction of people, places and events can coincide with current events and even help resolve pressing issues, but otherwise merely provide fantastic, whimsical, and sometimes even horrific visions.

There is no basis, many believe, upon which to conclude that dreams are anything but a biological function. In the end this may prove true, but what is the definition of biological function?

Dreams can be based in many cases on everyday life and populated with many familiar faces, people, and things. Dreams can also be fiction made up of the mind's fantasies, or the mind's nightmares.

People dream all the time, night and day, and they dream of all manner of things, places, people, horrors, fantasies, beautiful, ugly, science and art, the grand and the every-day...

Dreams are mostly forgotten, unrecorded. Some are not.

Some are remembered in great detail.

Then, night after night, do the collection of people's endless dreams encompass everything, every detail?

Every detail of everything that ever existed?

Maybe.

People dreamed of 9/11...

before and after.

People dream of disasters a lot...

before and after.

Can this be?

There has been documentation and research that indicates dreams can be precognitive – that is, they accurately reveal details of future events.

Are all dreams precognitive?

Probably not. Many are associated with current events in people's lives, some about the past.

Are dreams about knowledge; are they about facts, and details of actual events?

Are they accurate? Do they depict real life?

This one did, part of it.

The dream is a riddle, baffling to the awake mind, but with a sense that it is simple once understood.

The dream may never resolve, but a few pieces fit together.

First, somehow the knowing. And then, the importance of this knowing, the access to it and the knowing itself.

The dread angst repeats for several weeks after the dream, bludgeoning the logic of the awake mind into sense, or senselessness.

What is the truth?

The awake mind is nothing if not persistent.

Logic comes to bear when awake. But logic does not impede, for once, a remembrance.

It is simple, ordinary, a fact.

The dream is enormously detailed with familiar everyday places and things, places and things observed many times in the awake world.

Those details are accurate, a fact of the dream.

The dream is enormously detailed with unfamiliar everyday places and things, places and things never before observed in the awake world.

The logical awake mind comes to a startling conclusion.

If a good part of the dream is accurate, those of familiar everyday places and things, then the rest of the dream must be just as accurate. The feeling is strong, even awake, a sense of discovery, a sense of order, of what is.

But then the awake mind, after weeks of a foreboding sense of loss, seems to recall a conclusion. It is like a memory of a discovery that was not made, which does not make sense.

Everyone has these kinds of dreams is the conclusion.

All the time.

Awake in the night.

Asleep in the night.

Both intertwined.

Thoughts rattle from one to the other and back. Thoughts of impossible problems from the daylight.

Disorganized...

but organized...

a paradox.

Awareness flirts with disawareness...

color with black and white...

loud with soft.
And then, instantly, awareness.
The awake mind takes hold.
Not of the problems...
but of the solutions.
Not of yesterday...
but of tomorrow.
Not of the past...
but of the future.
Sleep on it.
It is a widely used expression, common and timeless.
But why?
Would it be common or timeless if there was nothing to it?
What does it really mean?
Why does it work?

In Malcolm Gladwell's book "Outliers," he describes a man named Christopher Langan as perhaps the smartest man in the country (his IQ the book indicates is 195 compared to the average of about 100 and Albert Einstein's at 150). In the book Christopher Langan is quoted:

"I found if I go to bed with a question on my mind, all I have to do is concentrate on the question before I go to sleep and I virtually always have the answer in the morning. Sometimes I realize what the answer is because I dreamt the answer and I can remember it. Other times I just feel the answer, and I start typing and the answer emerges onto the page."

While dreaming Friedrich August Kekule came up with the structure of Benzene, Dimitry Mendeleyev conjured up his final form of the periodic table of elements, Otto Loewi thought of the neuroscience experiment that won him a Nobel Prize in medicine, modern engineers Paul Horowitz and Alan Huang dreamed designs for the laser-telescope controls and laser computing respectively, Mary Shelley dreamed the two main scenes that became *Frankenstein*, Robert

Louis Stevenson did the same with *Dr. Jekyll and Mr, Hyde*, Ludwig van Beethoven, Paul McCartney, and Billy Joel all dreamed new tunes and Mahatma Ghandi's call for a nonviolent protest of British rule of India was inspired by a dream.["Answers in Your Dreams" by Deirdre Barrett, *Scientific American Mind*, November/December 2011, pg 28]

It seems this is common.

No need for a high IQ (except perhaps to process what is dreamt)

Perhaps it is universal.

Déjà vu is defined as, 'the illusion of already having experienced something actually being experienced for the first time.'

Everyone experiences déjà vu. It is nearly universal. But why?

Is this related to the expression 'sleep on it'?

Is it a similar phenomenon?

Could it be that you dreamed of a place and then when you actually go there, you feel as though you've been there before?

Could it be that you dreamed of an event happening, and when it does, you feel you already experienced it?

Could it be that you sensed something happened, a premonition, and then when it does, you feel you already experienced it?

Is this a knowing?

There are intrusions into the time between sleep and awake, the inbetween times.

Many.

To verify the unbelievable.

To transfer knowledge from sleep to awake, from dream to real.

To discover the true.

But the awake mind is rebuked.

The dream is of a Victorian home, palatially grand, almost gothic.

The details are stunning, 'realer than real.'

The omniscient viewpoint provides shivering insights to the interior.

Unimaginable appointments, artwork reminiscent of the French impressionists, perhaps Van Gogh. Ancient vases, intricate woodworking detail, gargoyle appointed furnishings.

All of it displayed in something more than human senses are capable of.

The intrusion comes as always from the awake world. What is it that is seen asleep in dreams? How is it seen in dreams? Where does the detail come from?

The awake mind arrests the dream, stops it cold and itself takes a hold of thought, wrestling it from sleep, from dreams. The dream is lucid.

Awake there is nothing of this detail, none of it ever encountered, never a wayward visit to Wikipedia, Google, nary a course or video on the Victorian or gothic eras.

Awake the mind asks does such a place exist, are the details real?

The Victorian and details dissolve, the smoke now of dreams, but the notion of it, the knowledge of the details, though not the details themselves, remain.

How much more detail might there be in dreams?

Is there detail of everything?

Where is it that a dreamer goes in dreams?

Everyone dreams.

Seven billion dreamers.

Do we dream of everything, every detail, every event?

Do we dream of everything in the Universe?

It is a recurring thought since the dream floating above the bay on a lake, and it does not go away.

Stephen King, notable author of horror and the macabre says about his short story "Everything's Eventual" that it wasn't really writing or creation but rather a discovery, as if finding the story on an archeological dig.

"... stories are artifacts: not really made things which we create (and can take credit for), but preexisting objects which we dig up." [Stephen King, "Everything's Eventual" 2002, Scribner]

Damon Lindelof and Carlton Cruise, writers and producers of the hit TV show "Lost" spoke in a season and series ending episode that recapped the entire 6 years of the show and aired on May 23, 2010. At the end of the recap show called "The Final Journey," Damon Lindelof described the creation of it; "This show doesn't feel like something we made. It feels like something that existed and we were able to tap into."

Thomas Edison, gifted mind, gifted inventor, perhaps the most gifted in modern times, describes his inventions as discovery. "People say I have created things," he wrote in 1911. "I have never created anything. I get impressions from the Universe at large and work them out, but I am only a plate on a record or a receiving apparatus... Thoughts are really impressions that we get from the outside."

Creative and imaginative writings for books and television, the invention of the light bulb, are these then discoveries of something pre-existing?

How do we make such discoveries, and from where?

Is this common in less grand everyday ways?

Is this common every day?

There is a wedding.

The father of the bride frets over the speech he must make.

What does he say about his daughter so beautiful in all of mind, body and soul, and whom he loves more than life itself?

Put in perspective, a small fret is really large.

A difficult task to sleep on...

A difficult task for a dream.

There is the latest issue of Wired magazine.

It is real, from the awake world.

On the cover there is a picture of a box wrapped in plain brown paper.

The plain brown paper is broken only by a large black question mark. The Wired magazine issue arrives about three weeks before the wedding.

The box it turns out is a gift to J. J. Abrams given to him by his grandfather when he was ten years old. It is unopened.

J. J. Abrams is a famous director and producer of television and film. To his credit are the television hit shows "Alias" and "Lost" and the blockbuster films "Star Trek" and "Super 8."

J. J. Abrams explains inside the issue that the contents of the gift are from a magic store, a subject he adored as a boy. Amazingly, a ten-year-old boy does not open it. A teenager does not open it. An adult of forty years does not open it.

For thirty years, J. J. Abrams does not know what is inside the box with a question mark, the one from his loving grandfather with gifts inside from a magic hobby store.

It is a mystery J. J. Abrams explains. If you open the box the mystery is gone.

Mystery it seems is important.

Mystery it seems is a destination.

Mystery is disappearing, Mr. Abrams reports in the article, what with sophisticated search at our fingertips and the answer to any question available instantly.

But not in dreams.

It is night.

Sleep comes easily, thankfully.

It is three days before the wedding and nothing much has been done about the father-of-the-bride speech.

In the wee hours there is a time between sleep and awake.

It is seconds...

It is an hour...

Mostly, it is a mystery.

In this time is discovered the speech.

All of it.

Nearly word by word.

Three days later it is delivered exactly as fashioned in the inbetween time.

There is a gift, the speech goes in part, underneath the sweetheart table of the bride and groom and only one person in the world knows what is inside.

It represents for now a mystery.

There is only one request that goes along with the gift. The bride and groom must not open it unless both agree to.

It is a journey that is begun, the creation of mystery. It is a journey of mystery. It is life.

And it is mystery too how this formulates in the night, asleep when the real world disappears to the solitude of a cacophony of shadows and a haunting and infinite knowledge.

There is foreknowledge of a graceful talk that could not have been delivered without it and could not be devised awake. Knowing something will happen, and knowing how it will happen provides confidence in the execution of it.

But is this a knowing? Is this foreknowledge? Could it ever be?

It is one thing to dream, another to be awake in a dream, to experience a lucid dream.

The inbetween times persist...

badgering, annoying, sleepless.

Discouraged but encouraged.

Answers exist here, perhaps all of them.

The mind works better here, maybe much better.

Naps, it is said improve creativity...

There is efficiency here entirely unobtainable elsewhere, a rapid placement of order to events that sometimes makes them understandable.

Is it discovery not of facts and events, but of understanding?

That there is importance, high importance, persists in the inbetween times.

Wake up in the middle of the night. Don't go back to sleep but don't wake up.

Stay there and cultivate this time.

Will the answers be there like ancient bones in an archeological dig?

When the answers come, with luck, are they of knowledge and events or are they understanding?

Is there a difference?

There are two faces. Vivid images.

The two people are old but not that old. Wrinkled skin here and there and grey hair. They are distinctive in portrait. These are two well known faces, worn from the pages of life.

Then the images of the faces dissolve fading, ghostlike, until in a mysterious cloud the ghosts materialize, replaced with two more images of the same faces.

This time they are younger, less wrinkles, no grey hair but the same two distinctive faces. The portraits are stunning and three dimensional far better than 3 D movie effects yet normal for dreams.

The ghostly cloud swarms again, and again the same two images of faces emerge this time still younger. They are at the pinnacle of their lives. Rich color and health resonate in the faces in the midst of conquering the world, ready and eager for their life's achievement.

They are, these two, achievers but still unremembered like so many visions in dreams. The images are friends or known celebrities but the dream does not keep track and the awake mind never recalls enough detail to tell who.

The ghostly ephemeral sequence continues and the two images are young adults, then teenagers, then young boys, then toddlers.

This is not fiction, it is a documentary. The images each time are real. They are the images of real people perfectly transcribed in portrait at each stage of their lives.

But this cannot be, the logic from the real awake world bellows. And yet the vivid imagery cannot be so easily declared fiction.

So how has the dream come to be?

There is a visit to the doctor during the day, a general practitioner who is old but not that old. An excellent doctor.

On the walls of his office are pictures of some events and of the doctor himself, some from various earlier stages of his life. He is youthful, vibrant in some of them, and older with wisdom and accomplishment in others.

The pictures on the wall do not lie. Pictures tell a story. That is why people take pictures, to remember stories. That is why movies are made.

The images in the dream that escaped to the real world are vivid, better than the pictures on the wall at the doctor's office. Can they lie?

If not they are the truth.

Again of something impossible to know.

It is like a computer program that either projects what someone will look like years from now or what they looked like years ago.

Did the dream play such a program? Can dreams play all such information?

There is logic, inescapable logic. It resonates from the awake mind, from the inbetween times, or dreams. Perhaps from the Universe.

It is a simple thought:

Events can be discovered from the past so they can also be discovered from the future.

A dilemma from the inbetween times.

It is a reminder there has been no resolution to the dream.

The simple statement oozes logic asleep but as the awake mind intrudes it is completely illogical. It is completely wrong.

A paradox, a mystery.

It goes on. If events are discovered from the past and future, what is the difference?

What is a reference point, a base point? Where do events begin and end?

How can the reference point be now, where everything that happened before is past and everything that happens after is future?

Fall further asleep, the conclusion comes instantly and it is discovery of fact.

There is no now...

No past...

No future...

No time...

Time does not exist.

The awake mind begins an eruption.

It objects with thunderous logic.

Everything on this planet progresses in time much like the progression from birth to death. Everything evolves, develops and devolves over time.

Nothing can happen without time.

Einstein's theory of relativity is about time. It comes from consideration of a clock. If you travel away from the hands of a big old clock tower like Big Ben in London and you travel faster and faster, as you approach the speed of light moving away from Big Ben, the hands of the clock slow down. As you reach the speed of light traveling away from the clock the light reflecting off the hands of the clock that enable you to see it cannot reach you since that light is limited to the speed of light as well. So the clock hands slow down as you approach the speed of light traveling away from it and stop altogether when you reach the speed of light.

Time stops.

This was proven in the laboratory by direct observation.

So, there is time.

The world is normal and functioning as you've always known it.

The awake mind is on a role subverting the whimsical fantasies of the dream world.

But there is doubt about the awake mind, about its logic. This does not seem to emanate from anywhere in the awake world.

At night, instantly there are thoughts to counter the awake mind, to silence the thunder of awake logic.

Time stops at the speed of light. Does that mean there is no time?

The awake mind operates in a box. The box is limited to a tiny part of the Universe called Earth.

Drift back toward sleep, toward the answers. The mystery needs freedom.

The resolution, at least part, is in dreams.

It comes then.

A message.

Logic, discovery, both?

It is overwhelming.

There is no now.

Light reaching the planet emanates from every corner of the Universe at every conceivable point in the past that ever existed. This light emanates from the stars.

Where do these thoughts come from? Certainly not an educated trained and experienced physicist.

So how is it valid?

In the night it is valid.

It is a knowing.

Sleep and awake are perfectly intermingled.

How can the vast Universe define now as the moment of these thoughts or the moment they are expressed when light from the Universe reaches Earth from every conceivable point in time? Why is Earth the center of the Universe of time?

That the vast Universe would always define now as current events on Earth, the past as everything that happens before current events on Earth and the future as everything that happens after current events on Earth is not logical.

So then, without a past, present, or future, without time, what is there instead?

Events.

It comes easy, it is simple but not original. Is it a deduction?

Theoretical physicists have postulated that the Universe is a bunch of events and time does not exist. There is no proof of this.

In commingled dreams and awake there is the notion of an encyclopedia.

What would the encyclopedia of the Universe be?

A description of all events.

Why would there be just a description of past events? Would such an earthly limit apply to the vast Universe?

Current encyclopedias chronicle events from this tiny part of the Universe, a part infinitesimally small.

An encyclopedia is alphabetical. Turn to any page and events of different periods in time are discussed, including some of the future (for instance the Sun will expire in a supernova 4 to 5 billion years from now).

To the encyclopedia there is no time, merely a description of all events.

The description of all events includes all observable events and also those events both past and future that can be deduced from observable events.

In the dream world, nearly asleep, it is obvious that with enough observable events all non-observable events can be accurately deduced.

Isn't this the scientific principle behind the Hubble Telescope? Perhaps all of science?

So if past events are discovered and deduced why wouldn't future events also be discovered and deduced?

To the Universe what differentiates a past from a future event?

In the dream there is no telling the time. It could have been yesterday, a week ago, months ago, years ago or it could have been any time, even those never experienced, never observed.

If dreams are of everything, why not tomorrow, next week, next month, next year, or years from now?

Why would there be a limit?

Awake, the mind wrests full control.

There is life to live, in a known observable environment. Things to do, things to arrange, a schedule to follow.

An arranged sequence of events.

At the very least an arranged sequence of events.

But for what purpose?

To manage in the work-a-day world.

To live...

To survive...

To thrive...

To pursue happiness...

To achieve.

So the mind is occupied, performing its function.

But yet, it intrudes on dreams, festers in the inbetween times, brooding about the mysteries, almost touching a seeming infinite vault of information.

Is it true what is discovered there?

Out of the shadows, the cobwebs of the night, might there be something useful, something to gain?

So the awake mind intrudes until there is no sleep and the connection passes...

A dreamer dreams. In the dreams there are wild fictions and yet recognizable truths. But who is to say what a wild fiction is? Who is to say anything is fiction? This time, this night the dream is lucid, at least part of it.

There is recognition of old things, very old things, some very old alive things. A woman is 119 years old. A manatee is 200 years old, some turtles, 300 years old.

Plant life is older still. Many trees live hundreds and even thousands of years. Some, the bristlecone pine, are 5,000 years old, the oldest living things.

We think.

But yet, the hills are older, the mountains are older, the continents, the seas, the planet, the sun, the stars. Thousands, millions, billions of years old.

It is stunning how old the Earth is, how old the Universe is.

It is stunning, even, how old some bristlecone pines are. At least from the perspective of a species that lives on average less than 80 years.

What is really stunning is how few years a human being lives. Eighty years is not even a microsecond in the life of the Universe or of the Sun. Yet humans appear to be the dominant life force on Earth.

Why such a short life span? It is so short, wisdom does not root. It is lost, it seems, as soon as gained because we die, generation after generation.

Reading a book is not the same as living.

Passing along wisdom in teaching and in books is not equivalent to blood, sweat and tears of actual living.

The notion of the dream is a wild fiction, but the dreamer almost understands.

There should not be a limit, a time limit, to the life of a human, to any biological life, especially one so infinitesimally short.

Humans should live to be thousands of years old.

Will humans live to be thousands of years old?

Why don't they?

Is it because in the beginning, lack of easy access to food especially in winter, disease, and bigger, stronger, faster predators prevented long life?

Culturally, generation after generation, such observation of reality was unavoidable. With the advance of medicine, of farming techniques, of food storage techniques, of greater protections from predators (mostly other humans), humans managed to live longer than 30 years on average.

Yet, after thousands of years, humans still do not even live to be 80 years old on average, even with modern medicine, modern technologies and the greatest era of peace on Earth in human history (death by war and human violence is no longer a significant factor in average age).

Is aging a disease, then, something that if cured would allow a thousand year life span?

No one seems to believe so, the lucid mind interjects. So aging must not be a disease.

Everyone accepts aging, and death is inescapable.

Everyone believes it, everyone observes it, so it must be (or is thus made to be) so.

There is no real reason that humans age the dream seems to say without saying it. There is no ingrained biological factor that requires aging, no rule of medicine or of classical physics that compels it.

Most obstacles to aging have long since passed, yet aging has not been studied very much at all.

Why?

From time to time, some people believe that they do not have a disease such as cancer even when diagnosed. This belief (observation) sometimes comes true (sometimes creates reality?). But this is crazy right?

There exists today a sort of mass hysteria about diseases like cancer and heart disease and the media reports all kinds of scary health statistics and stories. Does this contribute to, create, and authenticate a mass hysteria about disease.

Has everything been identified as causing cancer? Just about.

Is there any way, anymore, to believe that we will not contract some form of disease in our lifetime? If not cancer or heart disease, well then, we have to look forward to Alzheimer's or Parkinson's and while we can escape one or even a few of these, at least one is sure to get you.

Aging, therefore, does not exist, at least that would ever affect a life span. You will die of something else long before aging.

Such is the media message.

Such is the mass belief.

Such is the observation.

And thus, is reality created?

After all, quantum physics principles have been demonstrated. Part of what has been observed in laboratories, in studies of small particles, is that the act of observation creates reality or at the very least significantly affects reality.

Does the collective observation of aging make it real? Do mass transmissions (communications in all forms) from human to human (or to the interconnectedness of the Universe) make reality? If so, can such mass transmissions reverse the effect?

Science does not seem to believe that aging is something to treat, so aging is not treated, not studied, not researched, and not funded.

Not much anyway.

Can the medical sciences treat aging, stop it, even reverse it?

How would we know?

Collectively, the baby boom generation already has influenced how aging is being perceived. Seventy is the new forty and sixty is the new thirty. Many are buying into it and the media is reporting it.

The younger generation of boomer children are close to their parents, perhaps as never before, hoping and expecting their parents to live longer giving hope to the boomers that they can achieve a longer and higher quality of life. The combination of an expectation of the possibility of a longer high-quality life and the acceptance and strong desire of boomer children to harbor an older generation as an active part of their life, may rapidly replace a societal view of aging.

It would seem there is no scientific barrier preventing longer life.

Awake this is poppycock.

It is impossible.

Aging is a biological component of all of life.

Yet even the awake mind is transfixed by a question. Does believing make something real, especially the belief (observation) of a collective mass of humanity? Could this comport with quantum physics.

It is the Spring of 2002. The book "Flags of our Fathers" by James Bradley is a national best seller. It tells the story of James Bradley's father, and other fathers, who landed on the beach at Iwo Jima during World War II in a hail of mortal gunfire, grenades, and cannon fire. They conquered a bleak stone mountain held by the Japanese and raised a flag.

The flag raising arises from a most ugly, deadly and vile story of war, and somehow is made more heroic than actual events would attest to.

James' father, John, was a doctor in the Marines. He was by all accounts a hero on that day and the days surrounding the flag raising not only saving lives with his medical skills but also as a soldier. He, among six other Marines, raised the flag of the United States of America atop the tiny island of Iwo Jima just 600 miles from the Japanese mainland. He thus became part of, perhaps, the most famous photograph of all time. The photograph became the model for a far larger than life Marine memorial in Washington DC.

John Bradley never spoke of the war or Iwo Jima.

He never displayed the famous photograph or spoke of the awards he earned.

James, the son, and siblings, found memorabilia about the war and the battle of Iwo Jima in the attic of their father's home only after his death.

John Bradley only made one statement about the battle: "The real heroes of Iwo Jima were the guys who didn't come back."

There is an uncle, a father, a grandfather, who landed on the shores of Iwo Jima along with John Bradley, who also never spoke of the war. It seems there are few words and fewer still tales of any war from any of the participants, but why? James Bradley's book explains a lot about the silence but does not seem to explain the whole story.

In a dream, out of nowhere, the silence is explained. Why it is explained, why it becomes clear is unknown for the dreamer is not a veteran and has never been in battle. Perhaps it is a lifelong wonderment, shared by many, of what went on in the battles and trenches of World War II that arrested the reign of history's most evil leader.

No one can ever be a hero in a war, no one that has been in battle wants the label, no one wants the notice or recognition, no one wants any memory of it. There is nothing but horror.

But that is not all of it.

An Army Soldier, Marine, Navy man or woman can do many good things in a battle, in a war, including saving many lives. But for everyone in battle there are things that go wrong and comrades that you cannot save, no matter how many you do save. No matter how good you are, how brave or heroic, you will lose some of your comrades in arms, and in the case of Iwo Jima, many of them.

Because of the things that go wrong in the helter-skelter of battle, the utter chaos that no one can control, you do not and cannot save enough of your comrades because to save enough would be to save them all, and that in war is impossible.

It is the memory of those you cannot save, those that are not saved, those that did not return, that make the events of battle, of war, unspeakable.

You live with a haunting question: could you have done more, been a better, braver soldier, saved more good people, more friends who never came back with you?

No matter how good you were, the answer is always yes.

So, there is nothing to say, there are no heroes.

Awake, this thought settles as fact but goes unspoken. The receiver of the dream, of the thought, is not a veteran and has never been in the military let alone a battle or a war. The dreamer has no standing to say anything, especially a thought implanted from an unknown Universe.

An uncle has passed away some years who landed on the beach at Iwo Jima, survived the battle and the war, and became a prominent member of Tom Brokaw's "Greatest Generation." He helped rebuild a family, a small town in Pennsylvania and a sense of community...

But he never spoke of the battle or the war.

It is 2011 over drinks at an Irish bar in Washington DC. A Vietnam veteran makes reference to his long-ago hardened comrades in arms as if they might still be able to aptly discharge some unruly and drunken patrons. It is four rounds into happy hour, perhaps a good time, maybe the only time, to test a decade old theory. Nothing scientific, just one friend, one veteran and one war.

The dreamer asks the veteran if things go wrong in battle.

Many things, maybe nothing goes right.

The dreamer tells about his uncle and mentions casually that maybe his uncle could never speak about Iwo Jima not because of comrades he saves or because of the successes, but because of those he did not save; not the mistakes that anyone can make, but the moments of rescue that could not be because no one could save them all. Perhaps, it occurs to the dreamer, that his uncle had to kill in order not to be killed. That is war.

The Vietnam vet leans over his drink, drawn in, perhaps unnecessarily, to anguished memories, and looks up at the dreamer. He nods his head then and says "you're right," words that barely escape from breathless lungs.

Thirty five years after the war ended he had tears in his eyes thinking about what he could never say.

There is a waterfall.

It is spectacular.

On Lake George in the Adirondack Mountains of New York State.

It is vivid, breathtaking.

Like Bridle Veil falls in Yosemite, the water falls hundreds of feet off a granite cliff landing on golf cart sized boulders before bounding off in a white froth into the nearby lake water.

But this is a dream.

The falls does not exist otherwise.

From across the lake, the water floats down as if in slow motion.

It shatters apart on the boulders below shooting sparkles of spray many feet in all directions.

The broken water is a prism, thousands of them.

The sunlight skews against the prisms.

The rainbow is jaw dropping, the entire picture sublime.

But why?

Why this dream?

The dream is powerful, vivid, real, without reason.

Awake, the place of the falls is forest and cliff.

Forest and cliff should never be disappointing.

But it is now without the falls.

Should there be a falls?

Should New York State create one here where there are granite cliffs?

Was there once a falls here?

Of course.

Long ago Lake George was a hanging valley.

The daylight has answers, if but a few.

Much like Yosemite, Lake George was carved by giant glaciers in a past ice age that left granite cliffs stretching a thousand or more feet above the valley floor.

The high country provided water propelled by gravity down over the cliffs.

There were a bunch of water falls millions of years ago. Several more ice ages took away the high country and rounded the peaks.

Completely awake, was the dream about a real water fall that existed millions of years ago?

From the dream, somehow the falls is Native American.

The Lake always has been.

Native American culture abounds with legend and mystery. It is embedded with an ancient wisdom ascribed to the close connection of the culture with the land, with Mother Earth.

Such thoughts exist for most non-Native Americans without study or forethought as if ingrained in all American cultures.

The land vibrates, harmonizes with life upon it.

Native Americans have great insights into the natural world but little advantage is made of it.

Is it because any focus on the land has been lost by most Americans?

Most Americans would not dream of the land because they have been separated from it for generations.

So why the dream?

Is the dream not of the now, not of the future, but of the past?

Of beauty before people?

Is it a dream before the Pilgrims, before native peoples, before people at all?

On Google type in "a knowing."

The definition of the word knowing will appear multiple times in the search results.

"A Knowing" as it is used to mean knowing and understanding events never observed, and particularly future events, appears in a Native American website that itself appears at the very top of the search results.

This is odd because many books talk about the "Knowing" and a movie was made about one concept of "knowing."

But perhaps lack of precision in the search results is not.

At the top of the search results is found a teaching from Lee Standing Bear Moore recounted on the Manataka American Indian Council website:

"Knowing things does not come from learning things, it comes from remembering. Knowing comes from a place deep within us. The knowing is miraculous and instantaneous. It is old wisdom that remains inside us since the time of conception like tiny sleeping seeds, when it is needed by the world; the seeds sprout giving birth to what seems to be a new idea or solution to a new problem but, in fact the seed comes from a long line or root of knowing that is present at the time of our birth and long before...inside our souls and hearts.

"...We are not talking about information or knowledge. We are talking about the knowing – and that is a different matter.

"The knowing is something like intuition. That is, intuition is the ability to acquire knowledge without inference or the use of reason, but there is a vast well of knowing that goes beyond intuition. Like the knowing, intuition gives us ideas or thoughts that cannot be justified, but intuition is limited to the confines of human thought and reasoning and does not necessarily move beyond itself. However, it is said that intuition is a binder between human knowledge and higher spiritual knowledge and appears as flashes of insight and therefore moves freely between the earthly world and the spiritual world and the spiritual world is the same way of the knowing.

"Humans...possess at birth a knowing that transcends mere instinct...all without the necessity of training and education in these areas...

"...many humans never reacquire the ancient wisdom within. Some will experience only a quick flash of it at a critical moment in their lives...

"Moving past the veils of self (ego), social psychological and environmental (worldly) distraction requires intense focus... Many times this...can only be achieved in spiritual trance...

"...Only then may humans realize, take possession and utilize the knowing – after remembering – a process of calling up the ancient secrets of wisdom and placing them firmly in present consciousness.

The knowing cannot be learned from a book or other people...the knowing can only be remembered.

"Why do some humans live their entire lives without realizing the depth of the knowing within themselves?

"First it is a matter of distraction...by constant supervision of our parents, siblings, teachers, preachers, neighbors, friends and society in general...we are never taught how to remember.

"Second, it is a matter of orientation. We are never told there is this special knowing within us. We are never shown ways to access this knowledge. Instead our heads are crammed with rules, regulations, do and don'ts.

"...If we deny it [the knowing] exists, there is no way most people will ever begin to explore...the well of wisdom bequeathed by untold generations of ancestors...

"Third, it is also a matter of convenience and commitment. Most people are not motivated to discover things about themselves and the Universe if the road to get there is not well lit. When they discover there are no short cuts...along the way, they balk like a mule. They want to know the payoff in advance.

"How are we to know how to remember the knowing deep within us?

"There is a difference between what is learned and the knowing. Cognitive and recall skills can be practiced and improved over time...but the ways of recalling the knowing are different.

"The knowing does not speak English or any particular language. The knowing are images and movement that are not easily translated. The knowing is an understanding that flows like water or a cloud in the wind...

"...Knowing is the connection, the sacred connection between our inner-being and the Spirits of the Creator... the knowing happens by grace.

"The knowing is seeing the future. It is seeing the path and pointing the way. It is seeing time fold back on itself and reveal a disagreement that must be corrected. It is a deep wisdom that reflects the Universal agreements man has with all of creation. The knowing is immediately accepted 'as is' because it is truth."

"...When the knowing begins to flow from...deep within, there is an understanding that quietly flows upward in our consciousness and falls out in conversation."

Absent the silent thoughts of the night, the inbetween times, absent dreams, absent the thought that such things are common, even universal, the awake mind would ignore the teachings of Lee Standing Bear Moore.

Many generations have ignored the teachings of Lee Standing Bear Moore and other scholars from Native American Indian Nations.

Do you know anything unless other people believe what you know?

Does anyone know anything unless other people believe?

The Manataka American Indian Council website includes the "Dream Catcher" story:

"Long ago when the world was young an old Lakota spiritual leader was on a high mountain and had a vision. In this vision, Iktomi, the great trickster and teacher of wisdom, appeared in the form of a spider.

"Iktomi the spider picked up the elder's willow hoop which had feathers, horsehair, beads and offerings on it, and began to spin a web. He spoke to the elder about the cycles of life; how we begin our lives as infants, move on through childhood and onto adulthood.

"Finally, we go to the old age where we must be taken care of as infants, completing the cycle.

"But," Iktomi said as he continued to spin his web, "in each time of life there are many forces; some good and some bad. If you listen to the good forces, they will steer you in the right direction. But if you listen to the bad forces, they'll steer you in the wrong direction, and may hurt you. So these forces can help or can interfere with the harmony of nature."

"While the spider spoke he continued to spin his web.

"When Iktomi finished speaking of wisdom he gave the elder the web and said, "the web is a perfect circle with a hole in the center. Use the web to help your people reach their goals, make good use of their ideas, dreams and visions. If you believe in the Great Spirit the web will catch your good ideas and the bad ones will go through the hole."

"The elder passed on his vision to the people and now many Indian people hang a dream catcher above their bed to sift their dreams and visions. The good is captured in the web of life and carried with the people, but the evil in their dreams drop through the hole in the center of the web and are no longer a part of their lives. These are the words of a Lakota Elder which were given to Grandfather Hawk With Seven Eyes."

Many people, not just Native Americans, hang dream catchers above their beds.

Truth

The fabric of the night welcomes dreams, engages an unencumbered mind... is something there?

It is not nothing.

Is it everything?

A fabric thread throughout the Universe...

A portal...

A connection.

Gravity so exists, seemingly, weaving a pattern of force throughout the galaxies, throughout the Universe.

There is no description yet of its medium. Are there theories?

Of course.

No one describes the fabric of dreams. There do not appear to be any theories. Yet it could be more important than gravity, more essential to the workings of the Universe.

Is it something like dark matter, zero energy or gamma waves that some physicists claim permeates the entire Universe, something thoughts may ride upon like electrons in copper wire?

It is the inbetween times begging the medium to answer a question.

Why is politics untrue?

The dream ends abruptly not completing its story maybe not needing to. Maybe dreams never need to. But it does leave a thought, a truth about the real world.

It appears there are certain pockets of untruth enfolded in the structures of society. Used car sales people (perhaps many sales people of many products), real estate-speak, many media entities, customer service departments (partial truths are untruths), auto maintenance departments (how many times have repairs been made and parts replaced that did not need them?).

The list is long. Is it longer than the list of the true?

It is shameful how long such traditions have existed.

But the linchpin of untruth may be the legal profession for it is here, in a republic where rule is by law, that great minds spend great time "interpreting" everything without ever defining the truth. Does the client determine truth, or what the attorneys will argue as truth, or is it attorneys themselves?

Politics is a debate of law. It is undeniably dominated by the legal profession.

Great legal debate unfolds because great minds can always split one truth into shades of it slyly concealing self or organizational motivation. Is this then why the administration and Congress, the entire canvass of politics can be untrue?

Supreme Court Justice Robert Scalia is asked in an interview why so many American youth, often times the best and brightest, enter law school. This, he implies in the interview, is a waste of a good mind. Attorneys, he points out, do not really do anything. They do not make or build things, they do not create art, they do not discover new stars or planets or invent new products. All they do is interpret law.

"... the truth is hard to come by," a John Denver song hails. "And if I told the truth, that's not quite true." It is from the song "Some Days are Diamonds (Some Days are Stone)," words and music by John Denver, published in 1975 by the Tree Publishing Co. Inc.

The lyrics ring true in the fabric.

There are truths and untruths out there.

Do untruths seem true?

This makes the truth difficult to come by and an untruth easy.

Because untruth can work, the not true can succeed.

In politics untruth can work and the not true can succeed. At least for a time.

The true then is very difficult. But in dreams untruth is more apparent more readily seen. The true sometimes is clear, vivid.

Without truth governance is very difficult.

Everyone thinks what they believe is the truth. While this may be from time to time, it is likely that what we believe is often not true. History bears this out even when individually, and within human culture, this is completely unacknowledged. The untrue can never present themselves as untrue lest no one believe them, and thus objectivity is lost.

Within the fabric, the medium, the dark matter of the mind, within dreams there is knowledge, but knowledge is not an answer. An encyclopedia does not propose a resolution to world hunger.

Governance is difficult the fabric suggests. Very few governments in history have been successful.

With everyone believing different truths it is difficult to establish the true let alone lead.

Without truth it is difficult to lead a girl scout troop, a church committee, the PTA, a youth soccer organization, let alone a town, a county, a state, or a nation.

It is easy to understand the truth that there have been few governments that have run a nation even reasonably well.

The Roman Empire.

The British Empire.

The United States.

Perhaps the truth is that none have been run well, even reasonably well.

There have been no nations or states that have been run without corruption.

Why?

The truth is hard to come by, the untruth accessible, easy, but not really any part of the ether, of the fabric, of the soul of dreams, or of the inbetween times.

Does survival justify untruths?

How about thriving?

Where is the line drawn?

In this world, this ecology, the here and now, survival is the first pursuit.

Before nations, in hunter-gatherer cultures, truth was necessary for survival.

These berries are poison, these are nutritious. This is a bear's den, this is a habitable cave. With truth came life. With untruth came death.

If truth is how the world evolves, is untruth how the world devolves?

Untruth is manipulation of the truth for personal or organizational gain, but every step of that path is backward for the community of humans. Untruth disguises truth, distorts its recognition.

"Oh what a tangled web we weave, when we practice to deceive." Sir Walter Scott's quote is true. Might the tangled web be our world?

In a world of nations, individual survival is all but secure, food, clothing, shelter (noting of course many third world nations that need help).

In a world of nations, a nation's survival is not secure.

Survival becomes the first pursuit.

Does untruth pit one nation against another? Does it beg confrontation?

What one believes is true, someone else believes is untrue?

Is one code righteous, others not?

One nation is true, others with dissimilar truths are not?

Rule by majority, by law, by birthright, by manifest?

All good, or bad?

Have all been both?

If, then, it is not the style of government that is untrue, is it the leaders, the objectives, the strong incentive for survival that is untrue?

Or is it untruth that brings the untrue? Misinformation?

Is it also manipulation of what is believed is untrue. Would such manipulation exist in dreams, or be tethered amidst the threads of the fabric of the Universe, or found among rules that order the Universe?

Lucid now, the truth of this dream is the massive untruths of the awake world.

The dream ends, lucidity fleeting again with just a glimpse of what could be and answers not readily apparent. But there are questions.

It is not in dreams. It is not the inbetween times. It is not in daydreams.

It is a flash. It is awake reading a book, a novel, a work of fiction.

So what would flash in the untruth of a made-up story? Another untruth?

Yet isn't all fiction made of non-fiction? Perhaps festooned in dreams? Perhaps discovered in the fabric, the ether, the universal interconnectedness? Perhaps where the flash came from?

Perhaps fiction is not.

Perhaps the flash is not.

It is Michael Crichton's novel "State of Fear." It is 2004. Human caused climate change has become accepted science. A rapid accelerating pace of climate change is an accepted prediction. Catastrophic consequences is an accepted prediction. This acceptance is global.

But not quite, not to Michael Crichton.

The appendix to "State of Fear," a non-fiction "author's message," discusses the depth of research the author conducted to write the book and examine the facts of climate change. Michael Crichton concludes that the cause, extent of change and threat of climate change is largely unknown and unknowable. Studies, he concludes, that support a human caused catastrophic outcome are balanced against equally comprehensive studies that suggest the opposite. Some studies support the dawn of a new ice age.

Reading this message, there is a flash. It is simply that Michael Crichton got it right. But it is even more.

On the blank page that is the mind sometimes, with nothing going on there, where nothing should have been, there is a thought, delivered from nowhere.

There is no global warming, human caused or otherwise, no catastrophic climate change. Period.

Not yet anyway.

Without any basis, without mental exertion, this thought flashed. Not just there is no human caused climate change, but that there is no climate change, at least none not in a natural progression.

That cannot be, the awake mind asserts to itself, an internalized equivalent of talking to yourself.

Climate change is all over the media. Everyone and every organization is reporting on it.

The mind awake is overwhelmed by a flash that must be so wrong.

Michael Crichton's research was soundly rebuked by the major media and the idea of human caused catastrophic global climate change became entrenched as fact worldwide.

The pointed thinking was so intense that environmentalists and global warming pundits announced over and over that the scientific debate was over. It was time for action.

Later, even former pop rock star and founding member of the Beatles, Paul McCartney became vehement on the subject equating non-believers to those who do not believe in the Holocaust.

The flash, the mouse-click that raised the thought, is not the active mind but something deep below like Lee "Standing Bear" Moore described.

Though awake, it is not really from the awake mind. It is a reception, like receipt of a radio transmission, or as Thomas Edison described, a reception from the Universe.

The importance is to recognize it, discover its value, and use it.

Achieve.

But the message is wrong, the messenger unclear, the use of it unknowable.

Still, there it is, strong, a waxing gibbous.

The thought prevails, without research, without study, without any basis.

It becomes belief.

It becomes a certainty.

Not confidence.

Not ego.

Not false bravado or anything induced by alcohol or drugs.

Yet it is knowledge.

A certainty that something is true.

A certainty that something is not true.

Knowledge that may not be.

Then, in a wink, moments after the "understanding" that comes from Michael Crichton's words, there is still more.

Flashes happen all along, here and there, to everyone.

There is a flash sometimes on the true, sometimes on the untrue. Something "rings" true, something doesn't "sound" right. Something "feels" right, something "feels" wrong, like a puzzle piece that fits or doesn't.

Yet always there is uncertainty.

It is 2009.

Unwanted headlines trickle out in media reports of email exchanges between prominent climate scientists that seem to confirm that some of these scientists conspired to suppress data and studies that did not support human caused global warming and that did not support findings of cataclysmic climate change outcomes.

At first, the media suppresses reports of the scandal. A Tiger Woods scandal is reported unsparingly night and day for months, but something important, something that could shake the foundation of fundamental scientific knowledge, where the facts of science will shape world environmental policy for the foreseeable future, is not reported.

Al Gore's inconvenient truth might not be?

What is the truth and what is the untruth?

A difficult enterprise, this.

Is science mired in politics and agendas?

Is science mired in untruths?

Michael Crichton goes further with his comments.

He provides an example from history of global/viral think as a result of popularization and politicization of what he calls pseudo-science.

It is called eugenics, the study and practice of selective breeding applied to humans with the aim of improving the species.

Just the thought is unnerving.

Michael Crichton asserts that media hype based on pseudo-science convinced an entire generation of the facts of eugenics. Nazi Germany also subscribed to an ability to generate an improved Aryan race based on eugenics. Its plan, its actions, its acts of war and the Holocaust may have resulted in part from the belief and practice of pseudo-science. This science, however, was first developed and publicized in the United States. It was universally believed to be true.

The science of eugenics was eventually proven wrong.

Michael Crichton also mentions lysenkoism, a term that describes the manipulation or distortion of scientific process to reach a conclusion that would support an ideological bias related to social or political objectives.

He mentions all of these long before climate-gate was uncovered. Was this a knowing?

Is the flash?

If so, could nearly everybody be fooled? Again?

If so, nearly everyone is.

It is December of 2010. The magazine "the Atlantic" publishes an article entitled "The Danger of Cosmic Genius," by Kenneth Brower. It is about Freeman Dyson. The lead-in goes in part like this:

"In the range of his genius, Freeman Dyson is heir to Einstein – a visionary who has reshaped thinking in fields from math to astrophysics to medicine, and who has conceived nuclear-propelled spaceships designed to transport human colonists to distant planets. And yet on the matter of global warming, as an outspoken skeptic, he is dead wrong: wrong on the facts, wrong on the science. How could someone as smart as Dyson be so dumb about the environment? ..."

Freeman Dyson is a professor of Physics at the institute for advanced study at Princeton University in Princeton, New Jersey. The article states: "Among intelligent non-experts who have weighed in on climate change, Freeman Dyson has become, now that Michael Crichton is dead, perhaps our most prominent global warming skeptic."

On the "Charlie Rose" TV show in August 2009, the article points out that Freeman Dyson indicated "that he remained very skeptical about the dangers of global warming. He did not believe the pronouncements of the experts. He did not claim to be an expert himself, so he would not argue the details with anybody; he had not given much time to the issue and did not pretend to know the real answers, but what he knew for sure was the global warming experts did not know the answers, either... What he doubted was the models of the climatologists, and the grave consequences they predicted and the supposition that global warming is bad... Dyson argued that melting ice and the resulting rise in sea level is not cause for alarm. He said that the release of increasing volumes of carbon dioxide into the atmosphere is a very good thing, as it makes plants grow better. The important thing to remember, he said, is that the planet is warming mainly in places that are cold, and at night rather than during the day, so that the phenomenon is essentially making the climate more even, rather than just making everything hotter.

The article further mentions statements made by Freeman Dyson in a New York Book review article written by Dyson. In it he claims that the worldwide community of environmentalists, who are mostly not scientists, have adopted as an article of faith the belief that global warming is the greatest threat to the ecology of our planet. That is a mistake Dyson argues that distracts us from concentrating efforts on more serious problems."

The "Atlantic" article could be correct that Freeman Dyson is wrong about global warming. It is unlikely to be correct about Freeman Dyson being "dumb" about the environment. It could also be that Freeman Dyson is correct about global warming. What is interesting is that Freeman Dyson's arguments do not make the mainstream media while everyone has heard from Al Gore and many others on many mainstream media venues. This is odd because after the climate-gate exposure, 57% of Americans do not accept human-caused global warming as a major issue.

In the book "SuperFreakonomics" authors Steven D. Levitt and Stephen J. Dubner, describe a series of harrowing headlines and alarming media reports in the mid-1970s. "Some experts believe that mankind is on the threshold of a new pattern of adverse global climate change for which it is ill-prepared" the book quotes one New York Times article. The article includes the statement "this climate change poses a threat to the people of the world." News Week reported at the time that the National Academy of Sciences warned that climate change "would force economic and social adjustments on a worldwide scale" and that "climatologists are pessimistic that political leaders will take any positive action to compensate for the climate change or even to allay its effects."

The authors question who in their right mind wouldn't be scared of global warming in the presence of such alarming and apparent scientific evidence? Except that, they note in the book, "That's not what these scientists were talking about. These articles published in the mid-1970s, were predicting the effects of global *cooling*." The Earth was experiencing a period of time from 1945 through 1968 of cooling surface temperatures, greater snow cover and a decrease of 1.3% in the amount of sunshine hitting the United States from 1964 to 1972 according to "Freakonomics." The authors further note that Newsweek reported that though the temperature decline was small in absolute terms, the planet was already one sixth of the way toward the ice age average temperature.

So, it must be asked, were we really moving toward an ice age just a bit more than thirty years ago? And how can we be moving toward a global warming catastrophe today just a bit more than thirty years hence?

Is science confused?

"Freakonomics" goes on to chronicle the exploits of a group of genius scientists and idea makers led by Nathan Myhrvold who founded a firm called Intellectual Ventures (IV) based in Seattle. The firm, it is described, has a vast number of patents (20,000 and counting) and has developed proposed solutions to a number of difficult problems across many different fields. Indeed, Nathan Myhrvold, who founded IV, is described in the book as genius having earned numerous degrees in several scientific and mathematical disciplines and served as strategist and high advisor to Bill Gates while at Microsoft.

While the folks at IV agree that the Earth is getting warmer and they suspect that human activity likely has something to do with it, they also believe "that the standard global warming rhetoric in the media and political circles is oversimplified and exaggerated. In the book "SuperFreakonomics," Mr. Myhrvold indicates the purpose of Al Gore's film "An Inconvenient Truth" was "to scare the crap out of people." He further indicates in the book that while Al Gore "isn't technically lying" some of the catastrophic scenarios Mr. Gore describes, for instance the ocean rising to cover the state of Florida, "don't have any basis in physical reality in any reasonable time frame. No climate model shows them happening."

IV blames the scientific community as well indicating that the current generation of climate-prediction models are "enormously crude." Lowell Wood of IV states it more succinctly: "The climate models are crude in space and they're crude in time. So there's an enormous amount of natural phenomena they can't model. They can't do even giant storms like hurricanes."

So what explains the proliferation of climate models that predict the same dire circumstances which might lead to the conclusion that they are accurate? Mr. Woods, an astrophysicist, indicates that "Everybody turns their knobs so they aren't the outlier, because the outlying model is going to have difficulty getting funded." Thus, Woods further muses, the foibles and limitation of current funding mechanisms and scientific research should be taken account of when considering global ramifications of phenomena that cannot yet be accurately modeled. This would appear valuable for everything and not just global warming.

This group of genius scientists as described in "Freakonomics" believes that global warming might be a problem, especially if the rate of climate change remains high or quickens. If so, and particularly if there is about a 5% chance of some near term catastrophic rate of climate change, then something ought to be done about it. IV has thus developed several cost-effective proposed solutions that would take a few years to implement. They've created an insurance policy.

But the question remains, how can we ever be sure of the science enough to create a national policy or, much harder yet, a worldwide policy on climate change?

The debate of climate change rages on. The truth remains, as John Denver's lyrics suggest, hard to come by. Perhaps the truth of all of science is hard to come by.

Science is the truth and the untruth because throughout human history some of it has been correct and some of it has been wrong. This is not likely to change. It leads the way and gets in the way. It appears complete but never is. Sometimes it unclouds truth, sometimes it shrouds it.

Does the media turn group think into global think?

If so, how would anyone know?

Questions linger.

Is global-think prevalent?

If so, do bad ideas go viral? Perhaps more than good ones?

Does this limit ideas and muzzle achievement?

Does this disallow the truth?

No truth, no trust, no community, thus dysfunction.

The science behind the "no global warming" theory and its practitioners have voice again because of the interception of email, because of the internet.

Does the internet discipline science?

Does it allow the minority voice?

Do Google, Wikipedia, Facebook, Twitter, all the world wide web deter group think? Does all of it deter global think? Or does it create it?

Or cause it?

Does the internet facilitate every truth?

Does it facilitate every untruth?

How does anyone or any organization determine the good ideas from those that are not?

The truth is hard to come by.

So might ideas.

Education is truth. It is the teaching of it. In a dream this is thought of and believed unassailably true.

Teaching is how we formally define truth and then pass it along generation after generation.

What nations and cultures teach is generally thought to be true.

So everything that a nation is, that a culture is, that humans are, begins with education, with Teachers.

Without Teachers there is no education. Without education there is no truth. Without truth there is no nation, culture, or civilization.

Awake, this is not thought of, not part of a national conversation. Why?

Do Teachers define nations, cultures and civilizations? Does education rendered by Teachers define everything? Does it define the truth?

Boundaries

It comes from dreams, embroidered in a fabric that stretches infinitely across the Universe, the notion of us vs. them, of a line drawn in the sand, of ego, of governance. What is it all about?

Why such notions in the night? Can't it just be sleep?

But they do not shut up. They boil over to the awake world. The siren exists awake as in dreams. A call of something discovered, something sewn into the fabric, something transmitted between the dream and the fabric or the fabric and the dream, something like a connection.

There is truth in the connection, in the fabric, in the conscious. And the siren calls to find it, to enable it.

Awake, it becomes a conscious thought – there is nothing good about a line drawn in the sand, us versus them, winners that must make losers.

There is a "feel" of this truth almost like an emotion.

Yet, within the fabric there is something else. Always, there is something else, a deeper and deepening mystery, a deeper and deepening journey.

Awake, it translates to a thought that all of life is similar, far more similar than dissimilar.

Bears are more similar one to another than dissimilar.

Lions and tigers as well, oh my.

Dolphins, dogs, cats, all animals, mammals, plants, rocks, everything.

The Universe and its elements are more similar than dissimilar.

Humans are much more similar than dissimilar.

So why conflict? Why war?

Why is it difficult to govern?

Why two parties?

Why democrats and republicans?

Liberals and conservatives?

Why does one win and, in equal measure but opposite emotion, one loses.

None of this is achievement. Anyone can "feel" this truth.

Why is the truth hard to come by?

The American political system offers labels. Perhaps it is human to label and categorize.

Most folks are down the middle, more similar than dissimilar. Yet it feels like the Hatfield's and the McCoy's.

You're either for us or against us, good or evil, right or wrong.

The constitution does not dictate two parties, so why two parties? Why the Hatfield's and the McCoy's?

Take any issue.

Fix the sink.

The household gains because everyone uses the sink.

There is little communication between parents and teachers.

Get the PTA to establish more meet and greet times.

Everyone gains.

There is no youth girls basketball – create an organization to establish one. Everyone gains (yes even those families with all boys).

Take any issue and then:

Adhere to ethics,

Obey laws,

Weigh the public benefit,

Everyone gains, no one loses.

But what about nations with borders? Does that create winners and losers?

One nation has water aplenty, another is a dessert. One has oil, another does not. One has weapons, another, not so many.

What are the goals of governing a nation?

Survival is a big one.

Amongst all other nations, survival.

Amongst all other nations, general well being.

Amongst all other nations, improvement of the human condition.

All of these to make better everything within a national border.

And all of these at the pace and direction of each single government.

The emotional awake mind, ruled by belief, might be influenced by the unemotional fabric, the fabric of events, the collection perhaps of all events and all information. Discovering them is not simple, understanding harder still.

The question arises, why the line in the sand? Why nations. Why borders?

Why any construct?

Well, because governance is hard, the most difficult thing for humans to do effectively. We noted previously that just three governments have had an enduring reign in human history: the Roman Empire, the British Empire and the United States. But none without corruption, considerable corruption.

So, can humans possibly do the most good – one hundred percent for the public benefit. Without borders.

Is this a worldwide fable?

Fix the sink everywhere, everyone benefits.

Is it issues and not organizations, nations or politics?

Can such a focus be achieved?

No, it can't. Not without borders. Not without nations. Not yet. Humanity is not yet ready to shed its borders. Many would hold forth that the United States' constitutional republic, including the all-important Bill of Rights has been the most successful and benevolent government to the people it has governed in the history of human civilizations. People long for the freedom espoused by the Constitution and Bill of Rights and a system of laws that has been designed to protect freedom and the rights of the citizens governed. And yet the foundation of this form of government is continually challenged.

If you believe it has been the most successful government ever conceived, understand that it is a government of borders, not even one single government. There is one single federal government and fifty state governments. Plenty of borders. Why?

To protect against the loss of freedom and the loss of individual rights that history has proven a single government, without checks and balances on its leadership, has always taken away resulting in rampant corruption, tyranny and oppression of the governed people.

An eleven year old girl asks why is it that everything we know is wrong.

It is a simple question and the answer might be that the question is wrong.

But that is not the answer. The fact supposed by the question is logical and may even be absolutely right, but the answer to the question is difficult. What was known in physics, the science some would say of all, fifty years ago has proven to be, in large measure, incorrect today.

For that matter, the history of physics from Copernicus to Newton to Einstein has proven wrong time and again, with a worldview changing nearly every century and perhaps now every half century.

So how does an eleven year old ask such a perceptive question? How does she understand things are wrong or that things could be wrong, grown-up things?

These are questions of similar ilk to those of the night pressing the awake mind to understand what is wrong.

Not so much what is wrong but what is it that we don't see?

Is research, the influence of the trained and learned, and the experiences and thoughts from the awake world trustworthy?

Can it cloud things...?

Close in thinking...?

Is there need to expand beyond what is observed...?

Beyond the living and thriving function.

Beyond three dimensions, beyond local.

To what?

To thinking from somewhere else?

To knowledge from somewhere else?

From the corners of the Universe?

The dream is hypnotic, the call of the siren, compelling, addictive and unrelenting.

Asleep it is entirely true, these machinations from within, from without, from dreams, perhaps from the very fabric of the Universe.

Awake, what is true?

Some of it, most of it, all of it?

The mystery begs a solution evident in the night but not in the light.

It pops into the mind these thoughts of the Universe.

The strict order of science, its discipline, precise research, reliance on observation, myriad categories and branches...

Might it all be wrong?

The question an 11 year old asked.

Might it be set in motion by the ecology to which it associates...

and to the nature of life from which it is derived?

Is science made less aware, blind in some senses, unable to move beyond itself by the confines of human and Earthly ecologies?

Where science meets consciousness, thought, the fabric, the medium, the Universe's cloud, can one judge the other? Is something else altogether needed? A new worldview?

Perhaps a universal branch of science? Thought, consciousness, physics and all the other sciences, biology, chemistry, math...

Like an ebb and flow of tides, dreams and thoughts in the night sift as sand on a beach, looking for the right fit, searching out the truth, what is right, what is wrong, the awake mind investing itself of anything useful, of truths applicable to the awake world.

Is anyone thinking about combined science in the awake world? Is it all the same, just one science? Or does science not really exist?

Would such ideas be considered science?

Would anyone acknowledge such ideas?

Would anyone accept a new worldview?

Would anyone recognize a new worldview?

What is the tipping point for change in science?

Awake, what are these thoughts? Why these thoughts?

Einstein was an outcast. Schrödinger didn't even want to believe his own quantum theory which later was proven accurate and true. Faraday was labeled intellectually inferior but after his death his work became one of the pillars of physics. Galileo was prosecuted as a heretic.

History suggests all of these rejections were wrong. So, what are we wrong about today? I'll ask again, what are we wrong about today. How many ideas and people behind those ideas are we rejecting today, have we rejected over many years?

Is this how academia and the science communities have greeted all new ideas and thought throughout human history? Is it destiny? Is this human?

Achievement is the conception of ideas and their execution for universal benefit, yet humanity appears to consistently greet new ideas and thought with suspicion and outright contempt. Must this be overcome, not just for academia or science, but for everything?

If science has had it so wrong so often, has faith as well? There is no argument about religious beliefs. It is a birthright (in some countries, a human right).

If you believe from birth over a lifetime in your faith, when confronted by another what is made of it?

How do you not believe yours is right, yours is the only truth?

How could there possibly be objectivity? To do so would devalue a belief, a life, many lives and an entire culture standing many hundreds if not thousands of years.

Do the lifelong strong feelings of such beliefs lead to an inward belief, if not a public one, that one is right and others wrong or at least misplaced?

Must then sides be chosen?

Are beliefs right or wrong?

If beliefs are right or wrong, no matter the governance, background, creed, code, or religion, is the public benefit diminished?

It is a theme in dreams.

Everything is interconnected.

In dreams it is fact, awake a feeling, a universal feeling.

The awake mind, though, often invades the inbetween times, intermingles there with dreams.

Einstein is there, Schrödinger, Bohr, Faraday, Newton, Copernicus, Plato, even Hawking.

There is violent agreement among them, these great physicists and thinkers. In their day, they were right about the world, the Universe.

Their physics was correct.

But, it is not their day, not even Hawking's, and they are wrong about the world, about the Universe.

Their physics is incorrect.

Something is entirely missing, and it is vastly more important than what is not, a fact no different today than in Plato's time.

Is it interconnectedness? Is it entanglement?

Everything touches everything else.

In dreams this is fact so is this true awake?

But what is this interconnectedness, for what purpose, and how does it exist?

What even does it mean?

It is Stockholm Sweden. It is the occasion and ceremony of the awarding of the Nobel Prize for the sciences.

There are many of them awarded in almost countless disciplines of science.

This day for this ceremony an individual walks out on stage in front of the world community of scientists. The individual is to receive the Nobel Prize.

But the individual is not a scientist. Indeed, the individual has no degree at all, not in any scientific discipline, not in any discipline.

The individual is presented the Nobel Prize for science to an adoring crowd.

The individual is then handed the Nobel Prize for humanitarian contributions.

Finally, the individual is handed the Nobel Prize for literature.

The audience erupts in a thunderous roar. Then there is a hush as the individual begins to speak.

"I would like to thank the Nobel Prize organization for these honors, but more importantly for, just a few years ago, breaking down the barriers between the disciplines that the prizes had previously been limited to.

"I would not have received any honor in any of your previous categories. None of my work would even have been recognized.

"For you see I am but a humble mason by trade. I build things but I have no college degree.

"I had an idea of something. It was a big idea. But it would not appeal to those in physics, those in biology, those in chemistry, those in any specific science discipline. It would not be understood by those in humanitarian work and it was not at all rooted in literature.

"It was rooted in the connection of them all. The interconnectedness that you all have embraced.

"You have eliminated silos, and heavy-handed categorization. You have eliminated barriers to ideas.

"Finally, I am grateful for the internet. Social networking has finally grown up. It enabled my idea to find sponsors that enabled implementation for the greatest public benefit. This is the achievement, not mine but the world's."

Awake, this seems goofy. An award that spans nearly everything. But then, isn't that quantum physics?

Nonetheless there is logic. Modern quantum physics proves instantaneous entanglement. In a now famous experiment, physicists sent twin photons in opposite directions traveling the speed of light. They did so with the idea and capability of changing the polarization of one photon during flight. When they did change the polarization of one of the twin photons, the other changed at exactly the same time. How did they communicate? How did one twin photon change its polarization to match its twin instantaneously when traveling at the speed of light in opposite directions?

These questions remain unanswered. What it proved is that there is instantaneous influence and that Einstein's speed-of-light speed limit is not so hard and fast.

Perhaps everything is interconnected and has influence with everything else, and if so, then maybe there is just one science. Even awake this seems to "feel" right.

Are all barriers, all categorizations then just artificial, without basis?

Does this limit ideas and thinking?

Indeed, there are myriad branches of science even within a major field such as physics. Schools, colleges, and universities offer degrees in each of these hundreds of disciplines.

Who looks at all of it?

Together?

Is there an oversight board for all of science?

Scientists spend an entire career in one sub-discipline of a major field of science, sometimes on one project. Does this properly consider the details, facts, circumstances or influences of any other?

Can a seasoned chemist become a biologist?

Can a physicist?

A math professor?

Can a seasoned chemist become anything else?

Can a physicist?

A biologist?

A mathematician?

Perhaps they can, going back to college, starting at the bottom. But, who would do this?

Is this smoke-stacking? Is this a barrier? Why such boundaries?

Can a seasoned engineer of any discipline become an accountant? Can a seasoned accountant become an engineer?

Keep going and apply these questions to all career fields. What is the answer?

Is everything smoke-stacked?

Was this what the dream suggested?

In contrast to the Universe where everything may be interconnected, everything entangled?

Mark Zuckerberg studied both psychology and computer science at Harvard, an unusual combination. Facebook is psychology and computer science. Mark Zuckerberg says that he created Facebook because he understands psychology and computer science required to create a user-friendly social network.

History reveals similar connections have led to many achievements whether individually or through collaboration.

Does our culture foster collaboration and connections? How many may we be missing?

The dream continues a day later, a week later, perhaps a month or more, perhaps it is a memory of a dream from the past. There is no time in dreams.

Why is there no interchangeability of jobs? Can no one from one field contribute to another?

The dream suggests this is wrong, there should be interchangeability, and plenty of it.

The awake mind interjects this time not objecting. Social networks designed for problem solving provide evidence that viable solutions come from all sorts of individuals and usually not from those within the field in which the problem emanates.

Does this mean that social and organizational structures in place today are wrong? Does this suggest companies might operate a lot more efficiently and profitably if they mined such interconnectability from other fields?

It is a dream, but it is not a dream. Because it is a dream of the real world, not from the ether, the connection, or the fabric of consciousness, not from the Universe's cloud.

It is a dream of real things that exist today in the real world.

It is job prison.

And it is a nightmare.

The thought of changing jobs, finding a new one, is stressful. The problem of having to find one is worse.

Whether the current job is lost or lost meaning.

The job of finding a job is the most difficult job of all.

No one, it seems, in any discipline or field will hire someone who spent an entire career in another discipline or field.

That is given.

No experience in that discipline, no job.

No education in that discipline, no job.

Except, oddly, for CEOs, who interchange regularly across disciplines.

Why just CEOs?

So, we all seem stuck in smokestacks with uncrossable barriers and undividable boundaries.

Job opportunities are severely limited to, well, the same job.

Only somewhere else in some other company.

The more this happens the more this happens.

Locked into a career, forever.

Job prison.

How can this be?

Why should this be?

Forget about escape, except in dreams to the fabric of the Universe.

The fabric brings thoughts one of many nights. It brings a journey to the cloud.

It is not a problem of job prison...

Not a problem of the arts...

Nor of ideas, ethics, politics, government, academia, public benefit or of achievement.

It is one problem, not many, to which all are connected.

To which everything is connected.

It is a worldview problem.

The realization is inspirational.

But it is a dream, perhaps just a dream.

The awake mind is cumbersome and slow. It operates in real time, Earth time. It is consumed with real things and the art of living.

Dreams do not seem to be.

The mind is informed by a specific worldview based on, well, what people believe. What people believe is based on the ecology, culture, accepted science, and the way of human living bound by the local environment.

That is the message of the dream, or that is the message of the dreams, all of them.

Or is that merely the limitation of what the dreams say, confined by the worldview of the awake mind?

Perhaps observation (realization?) of the dream is limited by the observer. Perhaps all observation is limited by the observer in life as in dreams.

So, is the worldview constraining the observer's interpretation of dreams? Then does that worldview also constrain the observer's interpretation of life?

In one worldview, much of nature is complex, composed of many variables difficult to study in a collective ecology.

In another, there are few variables that define a simple structure that follow one simple elegant rule.

It is one problem.

Awake, the notion is unbelievable, unfounded, but oddly compelling, even fascinating with possibility.

Any worldview imposes a mindset on the awake mind, perhaps a limitation, a barrier or boundary.

There seems to be no focus in education, the media and certainly not government on these missteps of worldview. If there was, could we arrive at a clearer worldview, one with no boundaries or barriers?

The discovery and existence of infinite geologic structures and patterns defined by fractals is one example of many throughout human history.

Natural systems such as the branches of a tree or the coastline dividing ocean and continents were thought to be intricate and complex... random.

But the patterns of each, and of many more of Nature's structures, flow symmetrically and far more simply and elegantly than previously thought.

Fractals were not discovered until the 1970s, a relatively recent discovery for something so mathematically elegant and simple.

Fractals are a repeating geometric pattern such as a triangle or even a right angle or half of a rectangle or 2/3 of a triangle that are put together in different ways multiple times. They follow a few simple relationship rules.

Fractal patterns persist in nature describing such disparate geographies as a coastline, tree branches and roots, the human body's blood vessels and nerves or snowflakes. Fractals might also describe the jaggedness of mountain ranges, immense and small river systems and gorges (the Grand Tetons, the Mississippi river and its tributaries and the Grand Canyon for example), the color pattern of the wings of butterflies, the billows of storm clouds, the windblown pattern of Old Faithful...

Perhaps even the pattern of the solar system, stars, galaxies and of the Universe?

Humans didn't see it before.

What else don't we see?

It is one problem the dream said, to which all are connected.

Anyone can contribute to anything – and do it well.

There is one science, not many.

There is universal interconnectedness.

Everyone dreams and everything is dreamed.

There is one career, not many. Any separation must be overcome by interchange to untrench ideas, stir excitement and imagination.

All of such ideas, of such dreams, of such thoughts intersect.

And that is normal.

Half-awake in the inbetween times, it seems such notions are a strange coincidence or pattern of thought.

Yet it is clear there are no coincidences. The Universe, like cloud computing, is unlikely to harbor them.

The dreams from the fabric, from the collection, must then be of related matters at the very least, and probably of matters intricately entangled.

The dreams touch several scientifically accepted concepts, that of universal interconnectedness and its associated notion that conscious observation (perhaps even in dreams?) creates reality and induces universal influence (entanglement).

Moreover, the medium of dreams might well be consciousness whose "influence" may be instantaneous.

Awake there is excitement from thought. This is utterly fantastic or a super coincidence both of which are impossible, both of which are possible.

As physicists Bruce Rosenblum and Fred Kuttner state in their book, the "Quantum Enigma," "...the existence of quantum phenomena expands the scale of what is conceivable..."

Why does it take the discovery of quantum phenomena to expand the scale of what is conceivable to scientists? To anyone?

Why was there ever a limit?

Achievement is made of possibility not overlooked.

Achieveless are those who overlook possibility.

There is nothing "para" about paranormal or paraphenomena.

There may not be paraphenomena or phenomena at all. It may all be normal, just unexplained.

Cloaked by night and darkness inside a dream or thoughts surrounding a dream, it is safe to conclude these things are normal, but perhaps it still could never be accepted to say so awake and out loud.

Why?

Dreams are normal. They happen to everyone.

Wide interpretations of dreams have been postulated without a conclusion that such interpretations are ridiculous.

So, dreams with information impossible to verify, impossible for the dreamer to have observed, yet absolutely true, are possible.

Floating mysteriously above a bay on a lake in the Adirondack Mountains of New York State and observing physical things never before seen is possible and the truth of those things can be absolute.

And if so, everyone's dreams, parts of them anyway, accurately depicting everything, is also possible.

If demonstrated, this would confirm at least one form of the paranormal, extra sensory perception or ESP. ESP is the acquisition of information by some means other than the normal senses.

Maybe extra sensory perception is not really "extra."

Paranormal, parapsychology, paraphenomena, are scary terms almost unanimously perceived as outside of science. The long accepted public perception of them is that they are not true. Anyone who might believe otherwise is subject to ridicule and such beliefs disregarded particularly within the science community. For a respected scientist such belief can be career arresting.

The term "para" has an aura that no one of sane mind associates with, this despite the unanimous acceptance of quantum phenomena and the demonstration of instantaneous remote influence.

Author Larry Dossey's book "The Power of Premonitions," also discusses the landmark experiment, discussed earlier, where physicists sent twin photons in opposite directions at the speed of light with the ability during flight to change a characteristic of the photon called polarization. They demonstrated that when they changed the polarization of one photon the other photon moving in the opposite direction changed its polarization to match instantly. Mr. Dorsey further indicates that physicists believe that this and other similar experiments provide "unequivocal" evidence that non-local quantum connections permeate the entire Universe.

In Dean Radin's book entitled "Entangled Minds," the author provides an in-depth account of a multitude of studies conducted in the precise manner of accepted science that have demonstrated the existence of some para phenomena including, particularly, the ability to gain information from something other than the normal senses (ESP), precognition (the ability to know some future events) and remote viewing (the ability to send a signal to another person by mental means alone even across vast distances commonly known as telepathy). Moreover, these studies demonstrate that such experiences or abilities do not just happen to a few but are common human experiences.

Dean Radin, PhD, is a senior scientist at the Institute of Noetic Sciences (IONS) in Petaluma, California, and has led a number of these studies both for IONS and for other prestigious institutions such as Princeton University. The IONS website cites pioneering psychologist William James' (1902) description of the term noetic as:

"...states of insight into depths of truth unplumbed by the discursive intellect. They are illuminations, revelations, full of significance and importance, all inarticulate though they remain; and as a rule they carry with them a curious sense of authority."

The IONS website further describes the term noetic:

"There are several ways we can know the world around us. Science focuses on external observation and is grounded in objective evaluation, measurement, and experimentation. This is useful in increasing objectivity and reducing bias and inaccuracy as we interpret what we observe. But another way of knowing is subjective or internal, including gut feelings, intuition, and hunches—the way you know you love your children, for example, or experiences you have that cannot be explained or proven "rationally" but feel absolutely real. This way of knowing is what we call *noetic*."

If twin photons instantly "influence" the polarization of one another across a vast separation both traveling away from each other at the speed of light violating Einstein's strict universal speed limit, then fantastic non-intuitive discoveries are possible.

Mr. Dossey concludes that experiments documenting precognitive abilities are real and that the "local" theory that individual brains generate consciousness will be overturned to be replaced by a non-local view of the mind affirming that consciousness is "fundamental, omnipresent and eternal."

The view that consciousness was there in the first place and not created by the brain has been adopted by some physicists and philosophers. According to Mr. Dossey, physicist Freeman Dyson believes "the cosmos is suffused with consciousness from the grandest level to the most minute dimensions." Mr. Dossey also indicates that philosopher David J. Chalmers, Director of the Centre for Consciousness at the Australian national University in Canberra proposes that consciousness may be fundamental to the world.

This would seem to suggest a consciousness that the mind can tap into or is connected with, but science has no idea of the medium for transmittal.

Could dreams be a part of it?

Mr. Dossey points to extensive precognitive research and experimentation that purports to demonstrate that perceptions and thoughts can be sent across global distances between a sender and receiver (remote viewing) and that such thoughts can be perceived instantaneously.

Does this relate to the instantaneous remote influence of the twin photons?

The experiments also purport to demonstrate that thoughts transcend time since in some of the experiments receivers receive a thought before the sender sends it. The receiver is sensing a thought that will be made in the future.

Author Larry Dossey seems to relate the remote transmittal of thoughts – remote viewing – with consciousness. He indicates that physicists have demonstrated that non-locality is an inherent aspect of nature. The term non-locality means that a particular characteristic of our world is universal rather than something unique to a specific location such as planet Earth and the human environment and ecology on it.

Consciousness seems to have several meanings, one being awake as opposed to asleep or knocked unconscious and the other being this universal connectedness.

A new term for this latter more scientific view is perhaps needed based on what it does rather than any physical medium or transmission characteristic of it.

It is an interconnection and an entanglement. It is therefore intertangled, an intertanglement. The entire embroidery of the fabric across the vast Universe could thus be described as the intertangulate.

It is a dream, inside it a reception of thought. Non-locality by definition has no boundary.

Consciousness has no boundary.

The intertangulate has no boundary.

Belief

Ayoung boy of about eight years of age is on Christmas vacation visiting his grandparents with his family. He is bored, nothing to do, no one his age to do nothing with.

He is active, adventuresome, and grownup talk is unbearable. Outside there is snow on the ground this December in central Pennsylvania, new snow. But what to do with it?

The boy makes his way toward forbidden ground, the attic of an old motel that his grandparents run not too distant in character and charm from the Bate's motel. It is where they are staying, in one of the hotel rooms. On this vacation, the proprietors and the few guests are all sane.

The mere forbiddeness of the attic, as with most, is a grand invitation. It is above the third floor of an old clapboard house in modest repair that is attached to the hotel rooms.

The attic is a lifetime, several lifetimes, rendered in antiquities, period clothing, ancient war uniforms, old pictures of unknown people and even devices formally useful in the service of running a business. It will be a quarter century later before the boy will notice such things.

A young boy notices the handle of an old mechanical meat slicer and grabs it to spin the sharp cutting blade until he notices what he came for across the room. It will be another vacation, another year, this time without snow, that he will almost cut his finger off playing with the mechanical slicer.

His destiny, his fate this season lies across the attic floor where an old Lightning Glider sled hangs gathering dust and rust.

But he has sled before, so much so that he considers himself expert.

What catches his fancy is a pair of old wooden Broadmount skis with a label bearing the name "Lake Placid" and a date "1932" that he also will not notice for another quarter century. The manufacturer is Bredenberg Brothers, Inc. Champlain, NY. When the boy grows up the skis will hang on the great room wall of his Adirondack home.

The skis are beat up a bit with ancient 'bear trap' bindings, the kind that when snapped closed locks your foot in permanently.

Barely understanding what skiing is, the boy understands sliding, that these skis will help, and that there is a good-sized hill out back of the motel that extends to train tracks far below.

This day, this vacation, creates a lifetime fancy, the sport of snow skiing.

The boy tries out the skiis. Family lore records the event as a near death experience.

The skiis are twice the size of the boy, the bindings no fit at all, and the hill ungroomed with unskiable roots and rocks lining the way. At the end of the hill lay the train tracks that stop the last of his many runs. In the crash, his head bangs against the steel rails, his skiis sprawl perilously across the tracks. With his feet still attached to two tall oak boards, his body a ball of snow and ice, he manages to hop off the train tracks before the arrival of the next ten-ton locomotive. While it wasn't close, it wasn't comfortable.

Exhilaration, the boy would remember much later but that was not the description that others on that vacation used.

The boy fell in love with skiing and in turn caused his father to fall in love with skiing too.

The Christmas following, the boy's parents give him a brand new pair of skiis, his skiing destiny all but secured.

He learns to ski on a small hill near Hopewell Junction New York. All it has is a rope-toe but to a young boy it is like a trip to Stowe Vermont. A mere hill appears like the Swiss Alps.

The rope-toe is crude and set for the largest adult it may have to handle that day. The boy grabs it and suddenly he is airborne ... landing in a ruble, skis spewed across the path of the rope-toe an event that today would be aptly described as a yard sale. He is embarrassed but undeterred. Most of all he is determined.

After all what could be better than a lift up the mountain after spending every previous skiing venture walking laboriously back up every hill laden with heavy skiis and clothing.

There would be more trips to Hopewell Junction until one day the father plans a Saturday trip to Belleayre in the Catskill Mountains of New York State.

Belleayre has a chairlift, in fact several of them.

And mile long ski trails.

It is a real mountain, a real ski resort.

Here is the problem. The boy became ill Friday evening, quite ill, with a fever and flu-like symptoms.

If he remains ill, he would not be able to go; his father would not take him.

He grows even more ill as bedtime approaches.

He tells no one. He hides from his mother who like all mothers has a special nose for her children. He does not want her diagnosis because of its accuracy. Monday morning, yes for sure, please diagnose illness. But not Friday night, or Saturday morning before the thrill of yet a young lifetime can unfold.

At night he wills the illness away.

Nothing, absolutely nothing, will stop him from his first ski trip to a real ski mountain.

And nothing does.

He wakes up the next morning without any sign of illness.

He skies that day as never before and with boundless energy.

He has a great time with his father exploring and skiing all manner of mountain terrain and learning how to manage a chairlift.

And for the rest of his life, what touches the young boy the night before the trip to a real ski resort, or what he touches to make it happen, became a life-long love of snow skiing.

Was this a triumph of mind over matter, of mind over physics, of mind over body? Did the young boy believe enough that he changed the course of events? Did he observe something enough that it became reality?

Or, was it simply a 24 hour bug that ran its course in less time?

He was the only observer. No one else observed the illness. So, to every other possible observer, the boy was not ill.

Can anyone do this? Does everyone do this?

A young boy, age 10, visits the family doctor. It is a routine annual physical exam.

This time the doctor tells the young boy something that is not expected. Not anything terrible, just a prediction.

The boy, the doctor says, will grow to be five feet eight inches tall.

Now the doctor does not say the boy has cancer or will develop cancer or any other serious disease. In fact, the doctor confirms what the boy knows; that his health is exceptional.

But he says something very traumatic for a young boy.

That the boy will not grow up to be tall. He will grow up to be short.

The boy is dismayed and does not want to believe such a prediction. He decides that he does not want to be five foot eight inches tall. His father was five foot nine inches tall, and he did not like that.

He decides that he will be six feet tall.

That is it, a done deal. He will be six feet tall. End of discussion.

The boy does not think about it again and there are no more predictions.

It turns out the boy grows to be six feet tall dead even. His father was five feet nine inches tall, his grandfather five feet nine inches tall, his father's brother five feet eight inches tall and his mother five foot six inches tall.

But the boy turns out six feet tall, like he wanted, like he believed. It is belief for sure, but he also knew that the prediction was wrong.

Are belief and "knowing" entangled, each perhaps the predicate of the other?

There is a little girl of six or seven years of age. Things happen to her that she believes already happened. She 'feels' things she does and places she goes already occurred. When you are six or seven you assume such feelings are normal. Everyone must have them.

So, one day she asks her friends if such things happen to them, if they ever feel like they have been to a place before when they haven't or if events unfold that seem intimately familiar as if they have been experienced before. No one admits anything. It is too much to think about.

One of her friends finally listens intently as if she believes what the girl is experiencing could be real. She is a bit older and responds that the girl is experiencing déjà vu.

The girl has never heard the term before but is fascinated until her older friend indicates that such things are not real. Her older friend concludes rather forcefully that she does not believe the girl, that she is just making things up.

For a while, until things were summarily explained to her, the girl had déjà vu experiences quite frequently, perhaps not daily but weekly over the course of several years.

Growing older, and upon the council of her friends more and more about how such things can never be true, the notion of déjà vu fades away until such experiences stop forever.

In the dark, there is reflection, upon a moment, upon a day, upon a month, upon a year, upon a lifetime.

Do youth believe, and believe absolutely?

Do youth know what's coming? Is this a protection since few events are yet experienced?

Do youth influence reality much more than grown-ups since far less reality has yet been observed? With little actually observed is everything possible just by the observing (belief) of it?

Do young people grow out of this? Do they grow out of an ability? Do they grow out of a belief?

Do they observe more and more as they grow the realities already observed by grownups?

It is not a dream, not the inbetween times. It is not an awake dream or daydream.

It is a flash.

A flash of information, knowledge, a flash of knowing.

It comes fully awake. There is a sureness to it as if observing the notion of it, the thought of it, brings accuracy.

It is like suddenly knowing a passage from an encyclopedia.

Only the flash is of the future.

The flash is brought about by emotion, normally a suppressant of precognitive ability, but not this time.

It is brought about by a Hobson's choice – a boss that asks a subordinate not to take a vacation already paid for and not refundable because the employee is needed in the office at the particular time the employee scheduled vacation.

The boss indicates that he will honor whatever choice the employee makes but asks that the employee not take the vacation. The employee stands to lose thousands of dollars and credibility with good friends that the employee had long ago planned the vacation trip with.

There is no solution.

If the employee takes the vacation the boss will lose faith and confidence in the employee's dedication to getting the job done. If the employee cancels the vacation, he will never get that time back with his friends and may jeopardize future plans with those friends and even friendships themselves.

Emotion roils, like an argument might in weighing the pros and cons back and forth, except there are no words. There is no one from which to win the argument.

Taking a day to think about it the employee flashes, all emotion, anxious to take the vacation but knowing that such action ends any future at the company.

It happens in the heat of the moment. Perhaps it is supreme focus.

Perhaps the no solution scenario is unacceptable. A universe that hatches parables, paradoxes, unfathomable science, and dimensions is harbor as well to all of the solutions.

There is always a solution. That is a universal axiom.

In the end, there is no belief in a no-win situation. Where this comes from is unknown. Perhaps it is taken from fiction, from Captain Kirk of Star Trek who does not believe that a no-win situation can exist.

Perhaps if there is hard enough belief there is never a no-win situation. There is always a win-win possibility.

Perhaps the Universe makes it so.

If there is belief...

There is no observation of anything but the belief...

No observation of any reality but the win-win.

The flash is to accommodate the boss and forgo the vacation.

And, to do so knowing that the boss will withdraw the request for the employee to remain in the office.

The flash indicates the boss is testing loyalty, to the boss and to the job.

The flash is exactly what happens. The employee rests comfortably with the decision to forgo the vacation knowing the boss, upon the employee's acquiescence, will allow the vacation.

The win-win occurs with the boss happy, the employee happy and the work situation resolved.

Did the flash create belief?

Did belief deny any other reality (observation?)?

Was reality confined to just one possible circumstance?

Perhaps.

Daydream.

What is it?

Is it nonsense or laziness? Is it the mind and body needing repose, needing a sleep-like break during the busy day? Or perhaps a signal of something else?

Daydreaming about sports there is a flash. A flash within a daydream?

It is not a flash that the Lakers will beat the Celtics in game three of the NBA championship, or that any sports team will win any series or any single game. These must be a common type of pseudo-flash in light of the amount of betting on sports and the amount of money lost.

No, it is something else, something about the likes of Kobe Bryant and Michael Jordan before him and Tiger Woods.

There is something else entirely about world class athletes, something that has never been spoken about out loud.

And then because of the daydream, because of the flash, because everyone has them, the notion of world class athletes is not exclusive. It is a notion, a flash about everyone.

World class athletes appear to exhibit uncanny anticipation.

It is as if they know what will happen.

Do they?

Aroused by the flash, cognizant of it, the awake mind barges in, jarring aside the smoother, more comfortable thought patterns of the daydream.

World class athletes not only have well developed physical skills from genes, extreme training, and diet regimens, but they have extreme focus as well. It is this latter element that sets them much further apart from the ordinary everyday person.

Extreme focus allows a better understanding, a complete understanding of the playing field and the circumstances unfolding on it. You know what's coming, when it is coming, so you react more confidently and more boldly in response.

This confidence leads to belief, belief in performance, belief in an outcome.

There has always been something to belief.

Believe in something hard enough, work at it, and it can become true.

Does this kind of thing happen to everyone?

Is belief the kind of observation that creates reality akin to observers in the quantum realm?

Everyone who has participated in sports realizes that with greater self-belief, comes greater performance. But is there something more than mere confidence, practice, and natural born athletic ability?

Alex Rodriquez of the New York Yankees hit his 590th career home run and 20th career grand slam May 31, 2010 at Yankee Stadium.

Of the occasion he said, "I felt it coming for some reason. You kind of see things like that coming a little bit..."

Every single statement like that can be shrugged off and explained by the enormous talent of the athlete. Dig a little deeper and find that such a sentiment is a little bit more common and spread throughout the athletic universe.

Many sports psychologists suggest visualization techniques for improved success. Visualize hitting a home run and running around the bases to adoring fans, or something like that.

Visualize your success.

Believe in your success.

Believe in yourself.

Practice increases confidence and increases success.

But is there still more?

Roger Federer was, arguably, the best tennis player of his time. He won more major tournaments than anyone else. Records abound among his career statistics. He has excellent fundamentals, a great backhand, perhaps the best forehand the game has known, a great return of serve and exceptional volleying skills.

But if you saw him up close, and didn't know, you would not likely guess that he is a world class athlete. He is a bit taller than most at six feet one inch, but otherwise of modest build and leaner than most athletes. While he has exceptional tennis qualities and skills and exceptional tennis athletic ability, none of his skills or even athletic ability seem to be particularly better than other top ten players (except that he is exceptionally elegant and graceful, almost like a dancer or gymnast with a tennis racquet).

Yet he wins all the time. And he particularly wins the major tournaments or, at the very least for him, makes the final or semifinal. During the peak of his time in tennis none of the other players over a ten-year span came close to the consistency of excellence that Roger Federer displayed.

This extraordinary success does not seem explainable in ordinary terms. If you compare Roger Federer to Pete Sampras for instance, it would be difficult to conclude that Roger Federer is the better tennis player. Pete Sampras had a better serve, was even more athletic, had about the same or greater speed, had a better volley, an almost equivalent forehand and a bit less of a backhand than Roger Federer yet the latter has shattered the former's record for major championships and reigned as the number one tennis player in the world far longer than Pete Sampras over his career.

So, what explains the difference?

Watch Roger play. Closely.

He seems to move or react to his opponent before the opponent hits the ball.

Evidence?

As pointed out he is not particularly faster than several other top players, but when he faces the serve of the top servers in the game he is aced far less than other top players facing the same serves, much less than could be explained by the play of the game, physicals skills, or by chance.

It is as if he "sees" or "knows" where the opponent is going to serve. This is also true of Novak Djokovic who has eclipsed Federer in terms of major championship wins and is acknowledged as having the best return-of-serve of all time.

Unlike other players who seem to guess wrong the direction a big server is going to serve and therefore can only watch as the serve streaks pass them for an ace, Roger Federer's and Novak Djokovic's "guesses" seem to be correct much more frequently. Theye therefore move to put themselves in position to return the ball moments before the serve is actually struck so that they return the serve of even the fastest servers in the game far more frequently than other players.

Similarly, both players' ground strokes and movement to prepare to hit the ball seem to anticipate where their opponents will move to on the court or where their opponents will hit the ball. For instance, they will hit ground strokes for winners that wiz by barely a few feet from their world class opponents because they leaned or began to move in the other direction just as they are about to strike the ball. They therefore hit many more winners than their opponents and with fewer errors because they do not have to paste the corners with risky shots for winners when they can so easily "wrong foot" their opponents.

So Roger Federer and Novak Djokovic win many points not necessarily from superior or more consistent ground strokes or raw athletic ability but from a keen anticipation of what their opponents will do.

Do they actually "know" what their opponents will do before their opponents act? Or is this a superior ability to deduce what their opponents will do?

Is this the same thing that protects drivers in emergency situations where the driver "sees" an emergency developing on the roadway seconds before it happens. Is this equivalent to the more familiar and oft used description that time slows down in an emergency?

Does time slow down for the exceptional athlete during athletic contests?

Does this then exist for everyone, at least sometimes?

Do professional athletes find a way to cultivate this ability?

Is it the heat of battle that brings out a sharper focus for athletes?

Is it belief?

Or is it a "knowing" that spawns belief?

Perhaps an even greater example of a similar ability is that of Wayne Gretzky. Wayne Gretzky dominated professional hockey in a manner that no other athlete has ever dominated a major sport. Arguably, he dominated hockey greater than Michael Jordan or Bill Russell dominated professional basketball, greater than Babe Ruth or Lou Gehrig dominated baseball or greater than Tiger Woods dominated golf.

His statistics are startling. He holds nearly every scoring record in hockey, most by a wide margin including goals in a season, assists in a season, goals in a career, and assists in a career, total points in a season and total points in a career. Though retiring in 1999, he still holds 60 different scoring records (he held 61 when he retired).

It is the wide margin that is eye-popping. For instance, Gretzky holds the record for total points (goals plus assists) with 2,857 for his career during the regular season. The next best is a player named Mark Messier with 1,887 (at the time of this writing). Gretzky has more career assists than any other player has total points. He scored 894 regular season goals. The next best is 801 by Gordie Howe and the next best after him is Brett Hull with 741. He tallied 1,963 total assists and the next best is Ron Francis with 1,249. He owns the official record for most 100-point seasons with fifteen. The next best is ten. He owns the most 150 plus point seasons with 9 and the next best is 4. He is the only player to reach 200 points in a season and he did that four times.

If you met Wayne Gretzky at night in a dark street alley you would not feel threatened at all. If you met him in the daylight and did not know who he is, you would never guess that he not only is a world class athlete but perhaps the greatest player of a sport of all time.

Wayne Gretzky, in his playing days, was far from the fastest skater. He was far from the strongest, the tallest, the widest, or most muscular. His shooting ability was only a bit better than average. His slap-shot and even wrist shot were not among the top players in speed though perhaps each was far above average in accuracy.

What, then, set Wayne Gretzky apart, creating a human anomaly?

He had unparalleled stick handling ability. The puck stuck to his hockey stick like it was glue. He had unparalleled hand-eye coordination. And he could pass the puck; a particular forte.

Because of his stick handling ability, his ability to control the puck rushing up the ice, and his ability to pass the puck, what set Wayne Gretzky apart is that he rarely made a mistake with the puck. He rarely turned it over or made a pass that was intercepted, and when he made a pass, it got through to a teammate in a position to score a goal.

But none of this really explains Wayne Gretzky's excellence.

When he passed the puck, it seemed he knew not only where his teammates would be but exactly where defenders would be. It was as if the game for Wayne Gretzky was being played in slow motion, and for everyone else in real time. It was as if he had eyes in the back of his head. He knew his teammates, knew the pattern of skating developing in front of him and knew what each teammate and opponent would do in the next moments, enabling him to maneuver or skate around defenders or pass to a teammate who would come open for the pass after Wayne Gretzky had put the pass in motion.

He knew where his teammates and opponents would be on the ice like a mother knows what her children are up to even without looking. If you watched Wayne Gretzky play hockey it was apparent that he knew things that no one else did. The game would unfold before him in precisely the way he envisioned it, in a manner no other hockey player could anticipate.

Is this a form of "knowing"?

Do we all have this ability at some level, the great athlete just a bit more sharply?

Does it exist in the sciences, the arts, media, government, academia, politics?

Does it exist in everything?

Recall Thomas Edison, Stephen King, Lee Standing Bear Moore, and A-Rod.

Sleep on it, dream on it. It will come to you.

There is a young boy about to turn 18 years of age. He is about to graduate high school.

It is 1970. There is a draft to support a war in Vietnam. The government randomly chooses each day of the year one through three hundred and sixty-five, every birthday possibility. Young men are drafted in order of where their birthday falls on the government's list. It is well known that the military will need maybe half of all eligible young men so those with birthdays that fall in the first half of the draft list will likely be called into active duty. Those whose birthday falls in the second half will likely not have to serve.

The young boy's birthday is number 20 in the draft. He will be drafted into active service and likely see combat duty in Vietnam. His mother is worried sick about the selection. His brother is number 180 and will likely never be drafted.

The boy does not worry at all. He tells his mother the government will grant a four year deferment for those who attend college and he will be attending college. She knows this but it is not good enough.

After two and a half years of the deferment, the government announces abruptly that deferments will end after the current college year. The young man will have to report for duty following his junior year in college.

The young man still does not worry, but his mother is frantic, his father deeply concerned. The young man tells his mother, "I know I'm not going to be drafted, don't worry." But she is inconsolable.

The war ends in the spring of his junior year in college. The young man is never drafted.

"Que Sera Sera,
whatever will be will be,

the future's not ours to see,
Que Sera Sera,
what will be will be..."
"What will be will be" is a lyric from a famous Doris Day song written by Jay Livingston and Ray Evans. The question is could this be true?
What is fate?
What is destiny?
Along with the song, they are in this dream, a Universe of fates and destinies.
Is fate nothing more than dreams? Is it nothing more than actions? Is it nothing more than a life, a life prescribed, the prescription in dreams?
Perhaps.
Wherever the prescription is found, is fate and destiny based on a vision of the future, a premonition, precognition? Could both be discovery, an uncovering of what the truth is out of time?
Does this transform to belief? Is it belief or does it become so? If it becomes so, does belief transform into action?
What will be will be.

There is a very young couple about to give birth to their first child. It is an anxious moment for most when doubt creeps in about whether either will make a good father or mother. So on this occasion an older and wiser pediatrician determines to make older and wiser advice.

But what comes out is surprising and certainly not science.

"You know," the pediatrician says, "that you will be provided ample advice from your mother, your mother-in-law, your siblings, your doctor, your aunts and uncles, your friends. They all mean well, and it is best to shake your ahead, acknowledge the advice, and thank them for it.

"You can also read some very good books. Certainly Doctor Benjamin Spock's book is good. I know him well. He is a great pediatrician and a friend. There are other books as well.

"But let me tell you, all of the advice and even Dr. Spock's book, they are good and some may serve you well but, at the end of the day use your instincts, go with your gut. Everyone will mean well but you will know what is best for your child."

Perhaps the oddest thing was that, to a young and inexperienced couple, this seemed like sound advice.

There are many examples of instinct, of going with the gut, of belief, but what it all means is mystery.

Déjà vu is a remembrance of something you know without ever experiencing it before.

You know the phone will ring and who is on the line before it happens.

Why are your eyes closed every time a picture is taken of you? Because you 'know' when the flash will go off so you can protect your eyes? Yet the flash of the camera takes but a millisecond. It takes more than that to close your eyes. Do you know *ahead of time* when the flash will go off?

Time slows down in an emergency, or do you know what is about to occur moments before and can take preventive action to avoid an accident.

Why do you not get your tonsils out when the doctor tells you that the procedure will protect your health?

Why do you not get your wisdom teeth pulled when your dentist tells you the procedure will prevent major problems? Why is it that no major problems with the tonsils or wisdom teeth ever emerge?

Why do you decide, time after time, to decline having x-rays of your teeth taken when your dentist tells you they are needed to maintain the health of your teeth? After many years of declining such treatments, studies determine that such radiation even in small doses can cause harm.

At a dinner marking the end of a training class, the professor asks that each of his students say a few words about the class. When the professor announces that he will select the order of speakers randomly, how did a particular student know he would be picked first before the announcement even was made?

You know which foot you will use to step off a moving walkway long before the end of it without thinking (if you actually think about it you usually will get it wrong and may even stumble). It seems not to be calculation.

You stare at a person on the subway, on a bus or on a train and they look up directly at you as if they knew you were staring at them. You look up on a subway, a bus or a train to look right into the eyes of someone staring at you. Why did you look up? Because you knew they were staring at you?

Your mother puts you to bed with these words "Everything will be better in the morning," and you know what, she is right. Everything is almost always better in the morning. Is it because we "sleep on it?" Is it because the mind in sleep reassembles itself around a problem and the solution looks more apparent and easier to attain?

Is it because the mind in sleep reassembles reality? That would be impossible, right?

Awake, what are these things, how do they work? Is there a grand design? Or is it all just imagination, just a dream?

There is an experiment. In it, a group of elderly men are gathered for a retreat in New Hampshire. The men are divided into two groups. The first is asked to reminisce about their lives in 1959 aided by old issues of Life magazine, Jimmy Stewart films and discussions about people and events of the times such as Elvis Presley, Mickey Mantle and Fidel Castro. The second group is placed in the same environment but are told not just to talk about the good old days but to pretend they were young again and experiencing 1959 for the first time.

Both groups were found to benefit from the exercise. However, the second group who pretended to be young showed greater improvement on posture, strength, flexibility, vision, hearing, and intelligence. In other words, by the suggestion that they pretend to be young, they were.

Ellen Langer, Professor of Psychology at Harvard University, conducted this experiment 30 years ago. She appeared to have temporarily reversed the aging process simply by asking her subjects to believe they were younger.

In her recent book, "Counterclockwise: Mindful Health and the Power of Possibility," Ellen Langer discusses the placebo effect and the power of suggestion for medical treatments. Simple sugar pills prescribed by doctors for various ailments have proved as effective in many instances as drugs engineered for the treatment. The mere prescription by doctors of a placebo and the suggestion that it will work causes it to work.

In 2007, Langer told half of a group of hotel employees that their normal housekeeping duties met the Center for Disease Control's recommendations for exercise and a healthy lifestyle. She suggested nothing to the other half. After a month, the group that believed their work counted as exercise and believed they were healthier, indeed, really were healthier. This group lost two pounds each and improved their body-mass index and blood pressure without any change in physical activity.

Ellen Langer seems to suggest that patients that respond positively to the placebo effect and the power of suggestion should not be viewed as passive recipients of someone else's power of suggestion. If there is an expectation of healing that could be set in motion by a placebo, suggestion or the mindfulness of the individual themselves, then healing will occur.

Langer also sees that the language of medicine and of doctors can be damaging. For instance, a diagnosis that cancer is in remission cannot instill the belief that a patient is cured in quite the same way as a doctor who tells the patient that the cancer is cured. The latter may be a stretch of the absolute truth, but instilling belief from an authoritative doctor can make it happen. A lot of negativity from an authoritative doctor, Langer suggests, can lead a patient to believe there is no hope, which might be responsible for the death of some patients.

Ellen Langer suggests that authoritative attitude can instill belief and save lives.

If the few who might know of a diagnosis, maybe just the patient and the doctor for instance, believe in a diagnosis (outcome), then do they collectively observe the reality making it come true?

In the night, where dreams can explore such information against a backdrop of infinite information, it is far from an "aha" moment. It is more like a "so what" moment. Such information is not new, or more precise, or recently discovered. It has existed for all of existence.

It's just that, in the awake world, no one is using this information.

In the dream, Ellen Langer is right. It is not just what you know, it is what you do about what you know.

The giants of Native American Nations are not made prominent in American culture.

So none visit in dreams.

But in this dream is Lee Standing Bear Moore, his grandfather, his great grandfather, his son, his daughter, his grandson, his granddaughter.

The knowing is wrung out of most of us the dream goes as if pitched from a chorus of voices, from Lee Standing Bear Moore and all of his ancestors and descendants.

And then there is a single scholarly voice, quiet, thoughtful, pastoral.

"Why do some humans live their entire lives without realizing the depth of the knowing within themselves," Lee Standing Bear Moore is asking? "We are never taught how to remember... We are never told there is this special knowing within us. We are never shown ways to access this knowledge. Instead our heads are crammed with rules, regulations, dos and don'ts.

"...If we deny it [the knowing] exists, there is no way most people will ever begin to explore... the wisdom bequeathed by untold generations of ancestors."
Or the wisdom bequeathed by access in the night, in dreams to the entire of Universe of knowledge.

Ideas

Whose is the voice in the night, in dreams, in the inbetween times?
Self?
The Universe, the fabric, the ether?
Family or friends?
God?
There is clarity in the night. It derives from somewhere, from someone, from something.
There are dreams.
Then there is awake.
Between the two there is reflection.
And sometimes understanding.
Something perhaps the awake mind can settle on.
In the night it is clear.
An idea of ideas.
Everyone has an idea, a good one.
From time to time.
Everyone plugged into the Universe.
Everyone that dreams.
Everyone that connects.
Is it biological, built in, a survival ability?
Billions of ideas from everyone.
How many have been deployed? What is the guess? Across the globe how many ideas are implemented?
Thousands, maybe hundreds of thousands?
That is a large number but it sounds small.
What is the math?
Why would anyone dream of this stuff?

Awake, logic emerges. Good ole logic from the real world. Understandable but limited.

The math is small, very small. It is expected somehow from the dream but still very disappointing.

Start from the ideas anyone has over a lifetime. Confine thinking to ideas that might be implemented in some capacity to run a company, academia, or government. How many of those have been implemented?

For most it is none.

And yet everyone will have good ideas in a lifetime. One, several, maybe more.

Everyone dreams.

Yet very, very few of everyone's ideas get implemented.

A question from the night.

Whose ideas run the world?

Nothing has ever been accomplished but from an idea.

If so, whose ideas power achievement?

If not you, the boss?

The President?

The Google boys?

Bill Gates?

Warren Buffet?

The government?

Whose ideas power influential institutions?

Is it the employees? Ever get to tell the boss your great idea?

Mostly no.

Even if you do get a chance to tell the boss about the great idea is it implemented?

Not so much.

The dream, modified in the inbetween times, concludes that the top officials and advisors are guided by their own ideas.

Every organization?

The government?

Academia?

The administration?

Congress?

Does Congress have any ideas or does it merely react to them? Politically.

Does politics take over...? From ideas...? From action...? From achievement?

It would seem.

Do ideas that run the country come from the President and his cabinet?

Is this reality, observed by everyone, learned by the young?

Most of the time?

So, if everyone has one, or even more, really good ideas about important issues, issues of the human or planet condition, of running organizations or even making the world better, where are these billions of ideas?

It seems there are no statistics, no evidence, no numbers, no research about ideas, asleep in dreams or awake in the real world.

In the inbetween times the lack of qualitative evidence is dismissed. Proof is not needed in the depth of sleep-induced contemplation.

It is a dream, and inside it, an understanding, a grain of knowledge from a vast collection.

Millions, even billions of ideas are then never considered, never exposed to anyone or any institution that might implement them.

Are ideas then not sought out...

Not asked for...

Not mined?

Why?

Because they cannot be, there are too many?

Is this true, currently with a vast roiling media, a vast and instant internet and enormous and growing access?

Why shouldn't ideas be mined?

Why shouldn't ideas be asked for?

Why shouldn't ideas be motivated out from the corners of everyone's mind?

Human nature is the answer.

Everyone's idea is better than everyone else's.

Thus, top officials of any organization deploy their own ideas and dispose of everyone else's.

Is this true?

It is daylight and the awake mind is still doing the math. Logic, science, and math make solid sanctuaries from the darkness, from illogic, from something not based on these.

How many people's ideas run the world?

Arguably, maybe five or six people at the top of important organizations in business and government run the organization.

In government in the United States, it is the current administration, the President and cabinet, Congress and the Supreme Court. That's a few hundred. Double that for advisers to Congress and the Supreme Court and add in the leaders of the military. Double that again for State government leaders. Now count the leaders of the major businesses, media, and academic institutions. Throw in the influential celebrities and bloggers and other internet stars and what do you get? One hundred thousand? Maybe.

Multiply that times the number of influential nations across the globe (and some of these are dictatorships that control business and academia reducing severely the number of people whose ideas run those countries).

An educated guess becomes not more than half a million people's ideas run the world. These people do have good ideas, perhaps combined a million to at most 5 million between them.

Now consider seven billion people that on average will have or could have at least one great idea during their lifetimes. Doing the math by dividing 5 million by 7 billion you get 0.07% of all possible good ideas run the world.

Of all the ideas among seven billion people, only a few are sifted through to choose those that lead to action and, subsequently, to achievement among the businesses, academic institutions or governments across the globe.

Awake, it is perturbing. Such a small number of ideas run the world. Reflecting on the night and the genesis of this thinking, even the awake mind makes a startling but logical and mathematical conclusion.

Such a small number is statistically negligible.

Which means that few, if any, of the ideas that run the world are the best.

Negachieve.

What is music but an idea?

What is a novel but an idea?

What about a painting?

A song?

A sculpture?

A film?

A photograph?

What are all of the arts but a collection of ideas?

How many ideas must there be for the arts?

It is said that we all have a book in us.

Seven billion people.

Do we all have a song?

A movie?

A painting?

A sculpture?

It is dark. Nothing physical to see even with eyes wide open. But they are not and yet sight is clear, imagery vivid, the enactment of events precise.

All the events are here. The entire collection. The Universe's collection.

Why is this one playing, this particular one?

Seven billion people each with a book, perhaps a song, perhaps other of the arts.

All of the people that lived before going back as far as people existed. Tens of billions each with a book, perhaps a song, perhaps other arts.

That's a lot of art.

Where is it?

What movies do we see?

Some from Spielberg.

Some from Lucas.

Some from Cameron

Some from Bay

Some from the Cohen brothers.

Where are the rest of the ideas for great movies?

Why aren't they made?

Walk into a bookstore. What books are there to read?

Some from King.

Some from Grishom

Some from Patterson.

Some from Grafton.

Some fantasy (Harry Potter).

Some (a lot) vampire stories these days.

But somehow, the experience is always the same. The same authors, the same stories, the same book covers.

How come?

Why does James Cameron write the screenplays for his movies? Why does George Lucas?

If you have seen Titanic, Avatar, or any number of Star War movies, you know that there must be hundreds, thousands, perhaps hundreds of thousands of screen writers or potential screen writers more capable of writing dialogue than these two.

Ego.

The centric nature of humanity.

It is the dark of night and yet a human limitation is vivid.

Something that has not been understood awake.

Is humanity run by ego not humility? But such thoughts cannot escape the darkness of the night like an automobile's headlights blocked by a deepening fog.

Is it because they can that James Cameron and George Lucas and others write the screenplay for their own stories?

Is it to elevate art?

Is it to elevate ideas?

Is it to make the best possible film?

Is ego control?

Does control elevate ego?

Make no mistake, the authors and movie makers cited here are masterful, accomplished and even genius, and people flock to see their movies or read their books for a reason.

But are they the best?

Can they possibly be the best? Out of billions of ideas for stories, novels or film...

Out of billions of ideas for music, songs, symphonies, paintings, sculpture, photography, architecture, chemistry, physics, biology, cosmology and, well, everything?

Is anything in the real three dimensional world the best humanity can achieve?

Thoughts occur awake. Verification of sleep, of dreams, of the in-between times.

How about "American Idol?"

How about "America's Got Talent?"

Year after year these television shows uncover artistic talent, a lot of it, including world class acts. Perhaps even including the best or close to it that has yet been uncovered on the planet.

And they keep doing it, and the expectation is that they will keep on doing it.

Endless.

Before these shows, where were these artists?

Why did they not emerge previously with such enormous talent?

There is awareness of an enormous undiscovered talent, hidden, difficult to reach, infinite, like the Universe's collection hidden within the fabric.

How much talent?

Yet despite the discoveries, are such shows true? Do they speak of the uncovering of the arts and artists? Or are they simply a means to control content with the unwitting aid of the public?

Does the night translate to the day? Interpretations by the awake mind, searching for utility in the light of day.

Awake comes full bore. It is kind of like an abrupt movie ending that stuffs reality back into the mind, leaving just a fading memory and a haunting emptiness.

This time the notion is this: dreams and even thoughts from the inbetween times are multidimensional, perhaps more than three dimensions, perhaps, as physics might suggest, much more than three dimensional.

Do social structures, physics and the general confines that define the physical world restrict thinking and funnel billions of ideas into a relative few?

What motivates art and ideas?

The possibility of implementation?

The possibility of achievement?

Are these possibilities severely restricted?

Asleep there is endless possibility of achievement, awake sometimes just a withering memory.

No one has seen a million ideas in their lifetime let alone billions. Not a hundred thousand, not ten thousand; perhaps a thousand or two.

No one has enjoyed a million works of art in their lifetime let alone billions. Not a hundred thousand, not ten thousand; perhaps a few thousand.

No one has enjoyed the best of these billions of ideas or works of art.

No one has mined them, sought them out, sorted them, recorded them or otherwise accumulated them or even cared to do so.

Are there any cultures across the globe and down through history that even properly value ideas and works of art? For a few artists and works of art, perhaps.

Absent value, fewer ideas and art are created or even thought of, fewer are implemented, achievement is diminished and improvement of life's and the planet's condition muted.

Why should we suffer the ideas and art of the few, disaster after disaster – repeated mining disasters, the financial meltdown across the globe, Katrina, the Gulf oil spill, war after war, poverty, hunger, deplorable human rights violations?

Are the answers out there?

Is there motivation?

It comes from the dark shadows of the night such thoughts, bits of them, wildly scattered but somehow cohesive.

And they are assembled awake into something understandable, something related to events of the real world, something that makes sense. At least there is an attempt.

Awake, the three-dimensional world seems a maddening scramble for control.

Do the big publishers, recording studios and movie studios want control? Do they pursue control of content?

Of the arts, of the artists?

Is control limitation?

Does control constrain thought, ideas, and art? Does it constrain motivation and implementation of ideas and art?

Do social structures nurture control?

By the few?

Because of ego? (Is this the definition of politics?)

Is control ego?

Does control elevate ego?

Is control negative achievement?

How many have an issue with ego, their own or others'?

Is it everyone?

Negachieve.

In the night there is still more, another infinitesimally small grain.

What do the ideas that run the world focus on? What do the relatively small number of ideas that actually surface focus on?

What would be the ultimate focus of ideas?

Achievement.

What is the ultimate achievement?

Is it money, fame, fortune, happiness? Sometimes.

Would a better answer be improving the human condition, all of life's condition, and that of the planet's; improving Earth's community?

If so, how many of our ideas that run the world focus on this?

If so, how many people work on this?

The intruding and skeptical awake mind stays silent. It does not need statistics or research to 'know' the answers.

Most people work on a product or service produced or provided by an employer. Those products and services do improve the human condition.

Sometimes.

Proctor and Gamble make excellent household products that over many years have improved our daily lives. IBM created a workable computer, Google created workable search, each enabling availability and utility of knowledge.

Many companies work diligently to create a better work or living environment and do so earnestly and, recently, in respect of the environment.

These organizations and the work they do, the achievements they make are commendable.

But, are there larger achievements?

Perhaps the movement, recently, to go green?

Eliminate poverty and hunger.

Cure cancer and other medical ailments.

Stop terrorism.

Stop war.

Who works on these?

The government?

The Bill and Melinda Gates Foundation.

Bono.

Organizations that generally do not work for profit?

And the rest of us, the vast majority, work for profit.

Why?

There are dreams, more of them, of vivid detail. In them is fantastic knowledge.

But there is no message buried within them. No message of extreme importance, no further clues to the vastness of the collection, of the medium of passage of interconnectedness, of entanglement. None, at least, that are apparent, none that are received.

The dreams haunt of the importance, reminders of the hint of significant discovery, discovery that is only speculation, speculation that is incomplete.

That more machinations of the dream of floating above a bay on a Lake do not come, that greater insights to its meaning are not apparent, that further tapping into the collection for answers cannot be realized, is palpable even in the dark shadows that precede and follow sleep.

It is the awake world, perhaps, intruding into sleep, into dreams, into the inbetween times, consumed now with passion, with the possibility of illuminating discovery, bent on it.

But as emotion roils, dreams fade, and with them any connection to the infinite collection of information. A few snapshots come, but they are not revealing.

The precision and truth, the unemotion of the collection, the interconnected influence, the entanglement is fading for this engagement as if too much of a good thing is not.

It will not be this dream is the conclusion. The awake world has taken it over with its peculiar and specific logic perhaps only applicable to one specific ecology, one specific subset of the Universe.

So others come. They always do. A vast interconnectedness that will not end, a presence that persists day by day, infinitely, no matter our state alive or dust.

It is an energy, a fabric that emblankets us like air around the planet and space around the Universe, ever sending but difficult to receive. Does it power the mind? Does it power everything? Could it?

In the dream it is ideas.

What are they?

Are they signals? Programming? Some vast form of community? The help button?

Ideas are missing.

Some argue very well that ideas and great discoveries spring not from individual minds but from a collection.

Wired magazine [October, 2010] discussed a book by Steven Johnson entitled "Where Good Ideas Come From: The Natural History of Innovation." Wired describes that the author "draws on seven centuries of scientific and technological progress, from Gutenberg to GPS, to show what sorts of environments nurture ingenuity. He finds that great creative milieus, whether MIT or Los Alamos, New York City, or the World Wide Web, are like coral reefs – teeming, diverse colonies of creators who interact with and influence one another."

Wired also cites author Kevin Kelly's book "What Technology Wants," a history of some 50,000 years of technological advancement. Wired magazine describes author Kevin Kelly's assertion this way: "Technology... can be seen as a sort of autonomous life-form, with intrinsic goals toward which it gropes over the course of its long development." Those goals, he believes, "are much like the tendencies of biological life, which over time diversifies, specializes and (eventually) becomes more sentient."

Author Steven Johnson is quoted: "We share a fascination with the long history of simultaneous invention: cases where several people come up with the same ideas at almost exactly the same time. Calculus, the electrical battery, the telephone, the steam engine, the radio – all these groundbreaking innovations were hit upon by multiple inventors working in parallel with no knowledge of one another."

Author Kevin Kelly is quoted: "It's amazing that the myth of the lone genius has persisted for so long since simultaneous invention has always been the norm, not the exception. Anthropologists have shown that the same inventions tended to crop up in prehistory at roughly similar times, in the same order, among cultures on different continents that couldn't possibly have contacted one another."

Can history be so coincidental when components of the Universe are not likely to be? The Universe does not harbor, it would seem, coincidence.

Are pockets and teams of idea-makers rare?

How are they formed? How are they joined?

Seems like these authors point to an opportunity, a need unfulfilled. Perhaps the next great internet social enterprise; something more than trivia, more than cute videos of babies or pets, more than sharing electronic friendships.

Could such "ideas" be transmitted (instantaneous entanglement) once thought of or received; re-transmission? Could this explain why "simultaneous invention has always been the norm?"

What if the notion of simultaneous invention down through history described by authors Steven Johnson and Kevin Kelly is not coincidental...?

But from where do ideas originate? One person, several that others connect with, whole civilizations indigenous or extra-terrestrial, the interconnection, the fabric or God?

Is an idea simple communication with the fabric, with infinite entanglement that must contain everything?

In the vastness of sleep ideas are plentiful and available, like millions of bits of information made available for a computer in order for it to accurately calculate something.

Are these ideas, then, bits of information, the Universe's bits of information?

Ideas dreamed make sense. They fit like the seamless structure of a luxury automobile. One piece complements another to make up the whole.

So why not in the awake world?

Ideas are missing.

So many, the awake world looks like Swiss cheese, tattered with holes, and so much patchwork the original is lost.

The fabric of the Universe is replete with ideas making it whole. The awake world appears not.

Why?

Why are ideas easy asleep but not awake?

Asleep, ideas are accessible and they work. They are implemented, part of the harmonious whole.

Awake it is not simple.

So awake comes interfering with the reception, establishing but grainy snow that emblossoms a poorly tuned television picture.

Awake, the mind does not argue this time. There is wonder instead.

It is a finite culture, that of humans and of this planet. It is small and, yet, a mosaic. More pieces are undiscovered than discovered.

The dream partly awake, from the inbetween times, is understandable and appealing to the logic of the awake mind. The world at large is a mosaic, but perhaps fractally mosaic, functional in its dysfunctionality.

Life begs improvement, forward progress, but to where is unknown. For now, the dream suggests ideas, and the awake mind filters them to particular ideas, ideas to improve the human condition, ideas to improve the planet, to fill in the mosaic, if just one piece.

To take from dreams that tiny part that is understandable, that might even fit the small mosaic that is Earth's community, is akin to tuning a radio in a third world country with weak batteries that picks up mostly static and a few words here and there.

How are a few words deciphered?

Ideas are missing.

The culture is missing ideas, passing them by, unable to tune the few words amongst enormous static. Easy to do and thus, the likely continuing outcome. Perhaps this is intended?

What is the pattern?

Ideas are passing by.

They exist, are captured by an awake mind, but pass by, a piece of the mosaic that fits here and now but its fit is not recognized.

Ideas appear to be like pieces of information. They can be used and useful to anyone and not used and useless to anyone depending on receptivity, the dream seems to say.

The boss does not like an idea.

The University president does not like an idea.

The CEO does not like an idea.

The President of the United States does not like an idea.

Where do the ideas come from that these leaders like?

Where do the ideas come from that these leaders reject?

As before the sources are few.

Implementation is fewer still.

The dream circles back reverberating, itself an idea; an idea of ideas.

An idea to harvest ideas. Intriguing and addicting because of the mosaic, because of the importance, like a corner piece of a giant puzzle.

It has circled back, a piece of many dreams, a piece of them all.

It is not one dream, it is many. It is not many dreams, it is one. It is not a paradox this time.

If ideas are information, almost events themselves, unformed, not tuned, then ideas must originate everywhere, from every place and every thing.

Ideas may simply be a reading of events and putting together several, or taking away some, to form something that has never happened before, or something that could happen, or something that should happen.

Ideas make events, so make them good ones. They should be made well.

But what is a good idea and what is not?

Harvest all of the ideas, as many as can be the night seems to signal, and all of the good ones will be there.

Fractals were discovered in 1975. Previously science had the wrong idea. Science thought that the branches and roots of trees, the nervous system and blood vessels of humans, the fragmented and jagged nature of coastline or mountain ranges, and snowflakes among many other naturally occurring systems or structures are random.

They are not. Many, if not all, are fractals that obey unchanging but simple rules in their creation. (Is anything random?)

Richard Taylor is an artist and physicist who has become an expert in the study of fractals. He discovered that many of Jackson Pollock's 'drip' paintings were fractals themselves. Jackson Pollock developed a unique painting technique of pouring and dripping paint by three dimensional movement of his body on a canvass below or draped on a wall. He describes it this way:

"My painting does not come from the easel. I prefer to tack the unstretched canvas to the hard wall or the floor. I need the resistance of a hard surface. On the floor I am more at ease. I feel nearer, more part of the painting, since this way I can walk around it, work from the four sides and literally be *in* the painting.

"I continue to get further away from the usual painter's tools such as easel, palette, brushes, etc. I prefer sticks, trowels, knives and dripping fluid paint or a heavy impasto with sand, broken glass or other foreign matter added.

"When I am *in* my painting, I'm not aware of what I'm doing. It is only after a sort of 'get acquainted' period that I see what I have been about. I have no fear of making changes, destroying the image, etc., because the painting has a life of its own. I try to let it come through. It is only when I lose contact with the painting that the result is a mess. Otherwise there is pure harmony, an easy give and take, and the painting comes out well."

Was it Richard Taylor's scientific background that enabled him to recognize that he and a few friends accidentally created a fractal 'drip' painting by allowing the wind to move a makeshift pendulum dripping paint onto a canvas overnight in an art experiment at the Manchester School of Art in England? Or was it his interest in art that led to enrollment in the Manchester School of Art? The event is described in Matthew E. May's book "In Pursuit of Elegance."

Either way, the discovery appears to be a result of combining art with science. Without these combined interests, perhaps the amazing discovery that Jackson Pollock painted in fractals mirroring a most elemental and widespread structure of Mother Nature, "the fingerprint of nature" as fractals have been dubbed, may never have been discovered.

Make no mistake, Jackson Pollock had no idea what fractals were when he created his 'drip' paintings since he died in 1956 and science did not discover fractals until 1975. Richard Taylor's discovery was published in a 2002 article in *Scientific American* entitled "Order in Pollock's Chaos." His experiment in art that ended up replicating Pollock's painting style occurred in 1995.

In the book "In Pursuit of Elegance," Mathew E. May quotes Jackson Pollock: "My concerns are with the rhythms of nature. I am nature."

Indeed, Jackson Pollock recreated nature. His last painting to be sold at auction commanded $40 million.

How did he do it?

He was in tune with the rhythms of nature enabling an imitation of the three dimensional patterns of nature in his two dimensional paintings.

He had a special "feeling" about the style of his paintings and about his individual paintings.

Did he connect with the fabric?

Were his paintings a knowing, a prelude of knowledge of fractals? Did he and did we all understand about fractals long before science understood?

Are fractals a natural part of humans?

Were his paintings an idea of nature?

Or wasn't it an idea at all?

Could it have been a suggestion from the fabric, from the elemental structures that surrounded him, indeed, all of us, touching him in the way of discovery?

Jackson Pollock discovered fractals. Science would not prove it for more than 20 years.

Was this singular creation an original idea?

Can such things even exist?

Is it like painting a mountainscape, a coastline, any number of cloud formations, and thus, a replication of "the fingerprint of nature?"

Not original, but an exceptional rendition?

Is there no creativity, no singularity?

Is it all discovery of something pre-existing?

Is that all ideas are?

But there is more to this than an artist, and paintings, and "the fingerprints of nature." But it is not dreamed about and the awake mind seems feeble to undertake a deeper understanding.

Fractals seem so much more powerful than anything an awake mind has created, so much more infinite than a single mind, so much more infinite than the minds of all humanity, and perhaps that is it.

The awake mind has not created very much, not thought of anything much, not built anything much, not invented anything much except with the help of the mind asleep, a touch of the fabric, a connection to the ether.

Perhaps.

The mind awake seems alone, isolated, not interconnected, not entangled like an unconnected printer able to print anything but unable to figure out what to print.

The mind runs to what it knows, a similarity to fractals, one that is perhaps even more prevalent in nature but one that is inexplicably not taught in most schools. It is the "golden ratio," scientifically known as phi.

Phi is approximately the number 1.618 since it is an irrational number (the full number like pi has an infinite number of digits after the decimal point whose numbers never repeat). It is the ratio derived if you divide the length of one knuckle segment of your finger by the length of the next outermost segment of your finger. Your toes honor the same ration. It is also the length from your hip to your knee divided by the length of your knee to the bottom of your foot.

And so on, including dimensions of your face, dimensions found in many other life forms across the planet and dimensions found in natural systems and structures.

It appears, like fractals, to be part of the underlying rules that define the natural world.

Because of its prevalence and its aesthetics (that either we find pleasing or we are just used to) humans repeat it in architecture where it proves structural soundness, in music, and in other social and scientific endeavors.

Leonardo Da Vinci made famous the Vitruvian Man which was completed in 1490 and depicts the golden ratio applied to the dimensions of the human body. Da Vinci is famously thought to believe that the workings of the human body are analogous to the workings of the Universe including the dimensions of the human body that follow the golden ratio.

It appears that nature is made up of extremely intricate patterns that are constructed using simple and even elegant geometric rules.

Does the non-randomness of natural systems and structures previously thought random imply a design, an existing design? Does it imply that humans have done nothing but repeat what exists?

Does it imply non-locality?

Maybe.

David Bohm was a prominent physicist. He wrote "Wholeness and the Implicate Order" in 1980. He suggests an underlying order that is hidden and not obvious to the physical world. The explicit order is the physical, visible Universe and the implicit order is an underlying structure that may guide the explicit order together forming an Undivided wholeness.

David Bohm describes the relationship of the explicit and implicate order by referring to the flow of a stream:

"On this stream, one may see an ever-changing pattern of vortices, ripples, waves, splashes, etc., which evidently have no independent existence as such. Rather, they are abstracted from the flowing movement, arising and vanishing in the total process of the flow. Such transitory subsistence as may be possessed by these abstracted forms implies only a relative independence or autonomy of behavior, rather than absolutely independent existence as ultimate substances."

Awake, this thought seems like an abstraction itself, not a part of any reality. Yet the rules of fractals, the rule of phi, seem to suggest an underlying design, an "implicate order," as Bohm suggests.

Is there then an implication on free will?

Is everything determined in obeyance to an underlying implicate order?

Do we have independent existence? Is it limited based on rules that govern the Universe?

Are there rules not just for the physical Universe, but for thought, for the mind?

Are there rules for consciousness?

Are there only rules that must be adhered to, an implicate order that guides our actions, our thoughts, our souls?

Is humanity only a "transitory subsistence" within a sea of higher order?

Are our thoughts and ideas as well?

Maybe.

But maybe thought is on a different plain than the physical Universe.

Nourished by dreams, thought seems to be a power unto itself, related to the physical world but separate.

Perhaps, as David Bohm might describe, thought is part of an even higher implicate order than the rules of fractals and phi.

Achievement

The night, in dreams, in the inbetween times, is like a pinball machine playing in the mind. Awake and sleep intertwined, the steel metal ball of thought bouncing from one to the other, stealing kernels from the Universe's collection of knowledge, springing to the awake mind to test for utility and back again for understanding that is painfully obscure.

There is no certainty.

But there are thoughts.

Intriguing ones that evolve ideas.

Nothing motivates but achievement.

Can this be true?

The awake mind intrudes. What is the test of this?

How about money?

Money motivates, that is certain.

Money moves a free enterprise society like the United States. That is certain.

But in sleep, in the night, it is uncertain.

The motivation is to <u>achieve</u> money.

The achievement, however misplaced, is money.

For some, money is enough of a goal.

For some, success is enough of a goal.

For some, fame is enough of a goal.

For others, such achievement is hollow.

Family motivates. Everyone works for family. This is certain.

But not in the night.

Everyone works to secure the family, to secure self, to provide the basics, food, shelter, clothing, and then to provide comfort and happiness.

The achievement of a safe, healthy, comfortable and happy family is motivation.

If nothing motivates but achievement then motivation must vary in accord with the achievement.

The importance of the achievement being reached for determines the amount of effort.

There is enormous effort in achieving family security and happiness.

If this must be done working for an employer, that is what people will do no matter what that employment achieves.

So everyone concentrates on a job, doing it well, so that money can provide security for the family.

Hovering in the night, in dreams, almost like a halo, is the highest form of achievement. If, indeed, it is improving the human condition, all of life's condition, and that of the planet's, then, would any effort toward improving the community of Earth be the highest form of motivation?

Who is working on this?

How many are working on this?

Solving world hunger and poverty?

Ending terrorism?

Ending war?

Saving the planet?

Not many in comparison to all other more trivial pursuits?

Is achievement then muted?

Is motivation?

Family, or the achievement of a secure, happy family, is a great motivator. But then, nearly awake, and the thought not of the connection, or received from somewhere else, the family or self-motivation is the programming.

It is like the printer that can do more than print, but doesn't.

The mind can do more than survive and thrive for self and family, but doesn't.

At least often.

Why not both? And not just out of a desire to answer a higher calling, to put on that halo, but rather to establish that ambition as a cultural norm, a cultural pillar.

Work for family and self, and work improving Earth's community.

The highest achievement may require the highest motivation.
It is an endless connection, an endless collection.
The collection of the Universe's knowledge.
Is there access here in the night?

In the night, in dreams and the inbetween times, far from awake, there are rich ideas, innovation, unconfined thoughts, connections, even art, especially art.

Sometimes insights are gathered in the day, and in daylight dreams. Daydreams.

But awake there is a fading imagination of them and no implementation, disturbingly so.

The real three-dimensional awake world closes in, constricting thought and memory like rushing from Disneyland into a cardboard box.

The business of living, work, family, finances, emotion; these thoughts take over the mind.

Living occupies it.

Otherwise, humans may not succeed in the three-dimensional awake world.

Humanity might not thrive or even survive without such a focus.

There is a pile of money.
What to do with it?
Perhaps it is Apple's, Google's, Microsoft's or Berkshire Hathaway's.
What are <u>they</u> doing with it?
Does it belong to the shareholders', the employees', the customers', all the above?
Of these, which is most important?
What if the pile of money belonged to the Gates Foundation?
It is a dream of course.
A jigsaw puzzle this time as with many.
The dream ends frazzled, sketchy, fading.
It leaves a residue, a reverberating thought.

What if the pile of money that Apple owned, or Google, or Microsoft, or GE or IBM, what if it belonged to the Gates Foundation?

What then?

The Southwest flight attendant bade her passengers goodbye by thanking them... for her paycheck.

She is not paid by her employer, or by the shareholders, but rather by the customers.

So she herself announced publicly, an endearing sincere pronouncement.

The awake world gathers crushingly as it does most mornings. The night has left another thought, a premonition of sorts – that the era of greatest giving in human history has begun.

This, the flight attendant, and the pile of money is a riddle shrouded in the mysterious shadows of dismemory from the night.

But, the riddle persists, annoyingly, intriguingly.

The Bill and Melinda Gates Foundation has already been wrestling with what to do with about 30 billion donated to it by Warren Buffet, CEO of Berkshire Hathaway.

Mr. Buffet so graciously and generously has provided perhaps the greatest single donation to charity ever.

The greatest investor the world has ever known did not know how best to invest in charities. He did not know how to give his money away.

If Warren Buffet does not know how best to invest his charitable contributions, who does?

Perhaps the Gates Foundation, the most endowed charitable organization of all time?

These thoughts actually gather around the dream, exist because of it, awake thoughts the rational mind will use to set about solving the riddle.

Name the best organizations in the world.

Governments?

Private enterprise?

A charity?

An educational institution?

A religious institution?

While there are solid well run organizations in all of these disciplines, the likely choice for best organization is a well-run private enterprise such as Berkshire Hathaway, Apple, Google, Samsung or GE or a number of others.

So, why can't these organizations be charged with what best to do with a pile of money?

No reason, except they do not as a matter of course end up with a pile of cash from charitable contributions.

Why not?

The Southwest flight attendant understands that her paycheck comes from customers who are the general public. If she understands this, most employees can understand this.

The general public pays the salaries of most employees of any organization.

Does this create an obligation for the organization?

Sounds similar to expectations of charitable organizations.

Is it the same goal?

Do all organizations share the same goal?

These are rational awake thoughts now churning over and over, trying to undress and dress the dream at the same time. Riddles, puzzles, and mystery stimulate the mind awake.

Asleep, the mind needs no motivation.

If there is an obligation and similar goals why shouldn't the best companies (all companies?) devote management, innovation, resources and money for the public good?

But of course they do.

Some.

To some extent.

Why aren't all companies occupied with the public good?

One hundred percent of the time?

Nights later, weeks later, months later, in dreams, the inbetween times, it is clear, one hundred percent for the public good.

Optimize motivation.

Optimize achievement.

One hundred percent for Earth's community.

But awake, it is something else.

Far less for the public good.

Sometimes nothing.

Greed, power, corruption, ego

Self

Reality not dreams, (but are dreams real)?

Sometimes dreams are beautiful

What real should be.

Taunting, pointing, drawing a picture, a sharp one.

Reality or dreams.

Dreams or possibility.

Cultural aspirations?

Individuals volunteer money <u>and</u> time.

Does private enterprise?

Perhaps.

At least some.

Does private enterprise expend vast resources and money lobbying congress?

For what?

To solve world hunger...

Eliminate poverty...

Create better schools...

Eliminate war...

Provide a college education for all...

To enhance the environment...

To save wildlife...

To save the world?

There are examples of saving the world and examples of taking advantage of the world, of a position of power where responsibility meets profitability.

Drugs and the pharmacy companies, big pharma, are an example. Here they have a responsibility to research and develop pharmaceuticals that can provide an enormous service to the human population. To save lives, reduce pain and suffering, indeed, to produce products that extend life, make it better, richer is the highest form of achievement.

Some of big pharma achieve this some of the time.

And sometimes they fail spectacularly. And when they fail there seems to be motivation, high motivation, not to admit it, but to cover it up, to continue to produce a product that does not work or worse, one that causes, not prevents, health disorders.

Why?

Like a great hospital or exceptional medical team, big pharma has the ability to provide a halo-type of service and become a revered institution. But it has not. It is difficult for most people to name even one major drug company.

This appears to be because big pharma rushes drugs to market before value is demonstrated and before side effects are fully studied.

Many doctors fall into the trap of the easy diagnosis of prescribing drugs that have not been fully tested. So, too many drugs are prescribed and too many medical disorders are mis-treated, and too many drug companies are investigated for false claims.

Suddenly a halo-type service is exactly the opposite, the general public is in disarray as to who to trust.

A patient talks to a doctor about a sty in the eye that annoyingly will not go away. The doctor prescribes a drug in the form of an eye ointment and the patient picks up the ointment at the local pharmacy trusting the doctor, the pharmacy, and the drug company that it will remedy the medical disorder.

It does not. The sty is as healthy as ever.

The patient goes to another doctor. This doctor shakes his head in a knowing way as if plenty experienced in this matter. The doctor tells the patient not to use drugs. He tells the patient to boil up some strong tea, let it cool, soak a cloth in the tea and dab it onto the sty. There is a natural antibiotic in the tea the doctor says. It will work perfectly. And then the doctor says, perhaps playfully, perhaps not, don't tell anybody. It's as if the doctor might be chastised for prescribing something not accepted by the mainstream medical profession or as if he will be chastised or even blacklisted by big pharma.

The prescribed remedy works. If the doctor keeps doing this he will be revered by the patients but is there some sort of unspoken counter-culture that may prevent prescription of such commonly available remedies. Must everything go through the doctors and the drug companies?

If both the drug-prescribed treatment and the homespun treatment work, which should be recommended by doctors? What is actually recommended by doctors? Which doctor will the patient in this example return to? When this doctor does prescribe drugs, will this patient listen?

One hundred percent for the public good.

Apple may be the most successful tech company of all time. But what is the measure of success? If it is profits, Apple wins. If it is marketing, Apple wins. If it is designing cool gadgets, Apple wins.

What if the measure is the public good? Perhaps the outcome is quite different.

It is one thing for the Apple idea-makers to conceive of great ideas for gadgets that have a distinct utility to advantage people's lives and are perceived as cool. It is something else entirely to use such an established ability to toy with peoples fantasies. Microsoft does this with repeated versions of windows leaving the public bewildered whether or not there is need for the new version when the current version works well enough.

Does Apple create a new device, such as the iPad, and intentionally leave out desirable detail for later versions? As with Microsoft, it has happened for every release of all of its new products and so it is likely to be intentional. Is there any responsibility for Apple to produce the best product that it can for utility and advantage of its customer for as long as possible (customer service)? Is there something better than the besiege of version after version of a product marginally better than the previous. Is there something better than taking advantage of a branded cool image to wring as many sales as possible out of the public.

"If you don't have an iPhone, well you don't have an iPhone," Apple's tag line goes creating a public desire, as best it can, to be cool which means buying the newest version of Apple products as soon as they are released time after time. It is akin to the Nike sneakers that were, and still are, cool but cause an undue demand especially among folks who cannot afford $150 sneakers or $600 iPads.

It can be argued that, no matter the barrage of marketing and advertisements, it is the public's choice whether or not to buy, and it can be argued that a company should not abuse its popularity by making the public, its customers, buy stuff that is not needed. There is the green concern that we produce too much throwaway stuff such as multiple versions of the iPhone year after year just because a company like Apple, or for that matter Samsung, can.

Suppose Apple didn't advertise. Suppose no one advertised. Could a company bring out a release of a product every year or two?

What is the public good? What is Earth's community good? Are Apple, Microsoft and Nike achieving it?

Patagonia, the outdoor gear and apparel company, sells more by encouraging customers to buy less.

What?

'Fast Company,' the business magazine, quotes Patagonia founder and noted environmentalist, Yvon Chouinard, "Every time we've made a decision to do the right thing it ends up being good business."

The company made $500 million in sales in 2011 growing some 30% while setting the bar for sustainability. In the March 2012 issue 'Fast Company' names Patagonia the 14th out of 50 most innovative companies in the world for, among other things, its 'Common Threads' initiative. Patagonia got 24,000 people to pledge to buy less and buy used and then partnered with eBay to make it easy for people to make the pledge and buy and sell gear to and from one another.

There are others like Patagonia. The numbers are growing, the private sector business culture evolving. Perhaps the private sector will evolve a greater concern for the public good than government. This could reduce the tax burden sharply. Perhaps it should. One hundred percent for the good of Earth's community.

There is money in good ideas, perhaps great money in great ideas.

Most (all) of them?

Money is never made but from an idea.

Big money is never made but from a big idea, or an idea gone big.

Big ideas are being exposed on the internet more and more every day.

Ideas of public policy emerge.

Good ideas.

Big ideas.

Big money possibilities.

Should private enterprise immerse itself in ideas of public policy?

Or is that job solely the government's?

Google.org, Google's non-profit arm, has busied itself solving the energy problem concentrating on development of reasonably priced sustainable and renewable energy supplies.

This is not solely altruistic. Its large computer farms consume quantum amounts of electricity, increasingly so as it drives an impossibly powerful and ever enlarging computer cloud. It also focuses on power supplies to backup the existing electric grid and provide nearly 100% electric reliability to its burgeoning cloud, something the larger grid needs to improve upon in order not to damage irreparably either micro-economies who cannot afford their own backup supplies or the larger global economy.

Is there a sense of responsibility at play in Google's decision to create Google.org?

Maybe.

Is there an obligation to the public good, expanding a public good it already achieves by producing a useful search tool and other software innovations that are free for the public to use?

It is evident that Google believes public policy on energy set forth by Congress, the current administration, or government in general will not, at least quickly, produce reasonably priced sustainable energy.

"Ask not what your country can do for you, but what you can do for your country." Did President John F. Kennedy mean companies as well, and does country mean Earth's community? Does it now?

Should there be more efforts like Google's from private enterprise?

Has Google maintained such an attitude? Has it digressed to something else?

Does philanthropy have to be non-profit? Based on Patagonia's example, it does not.

It is the dreams, the inbetween times, the sweet spot between awake and asleep that form the words Public Policy 2.0.

Sweet dreams

Is this just an expression?

Will a non-profit arm work as hard, be as effective, be as efficient, garner as much support and resources needed to do the job as a for-profit enterprize?

Companies have formed solely around green.

Small ideas and big ideas to foster a green culture.

To save the planet.

For profit.

One hundred percent for the public good? Not yet, but a good idea.

What about other public policy goals?

Not yet.

Where do the dreams point? Tauntingly.

So much possibility, so much potential achievement.

What is the public good?

What is all of life's good?

What is the planet's good?

There may be some different opinions in this regard.

What does improving the human condition mean?

What does saving the world mean?

Can corporations improve profits by improving the human condition? Consider that an improved public condition creates happily employed customers making money themselves who can afford to buy products and services.

Sounds obvious.

But can for-profit organizations effectively address public policy issues?

Is there a will?

The dream seemed to provide a knowing. It will be, but there is emotion – a desire that it should be. Perhaps this is solely because of the spectacular failure of government to be to be guided by the best interests of its customers – all United States citizens – and its spectacular failure to be efficient.

There is no emotion in most dreams, little in the inbetween times until dreams invoke the awake mind's emotion, the awake mind's reaction to a dream, the awake mind's injection of emotion into a dream formulating a bad dream or a nightmare.

A dream not of the fabric but of the fears of the awake world.

Dreams occur every night and the vast majority of them do not come to this.

There is always emotion awake.

Why?

Without emotion, thoughts seem clear, visible, obvious, uncluttered, the view of emotion obvious.

Thoughts seem like the truth, like fact.

Driving down a busy road not thinking about anything in particular (day dreaming?) a thought pops into the mind. It comes from nowhere, a reception, a flash. All of the stoplights between this moment and arrival at home will be green, nine in total, an almost impossible eventuality.

The thought seems clear, unfettered like a clear radio transmission sent and received.

On this trip each of the stoplights is green rendering the thought important, the trip fast and uncluttered.

There is desire for green lights, an emotion.

Green lights reduce impatience and frustration, both emotions.

Try to repeat the knowing of green lights on subsequent trips.

Doesn't happen perhaps because of the astronomical odds against all green lights.

But it begs the question does pre-thought or conscious thought prevent a "knowing?" Is it better that something pop in the mind rather than making any conscious effort to think about it? Is "knowing" frustrated by an awake mind, by emotion, is it blocked? If so, does this support the notion that such things are receptions from outside like Thomas Edison described, receptions possibly from the Universe at large?

Is the connection to the fabric, to the collection, blocked by emotion?

After all, the casinos in Las Vegas make money and never go bankrupt by any common ability to "know" the draw of playing cards or roll of the dice. Lucky streaks happen all the time but, in a flash (of emotion?), they are gone. Of course, the odds of such games of chance are set greatly in favor of the house and therefore take account of the influence of minds that might "know" or even influence the results.

Do emotions help in the real, awake world? Star Trek would have you believe so with both the computer-like characters called "Spock" and then "Data" used to contrast life with and without emotion.

Is a hunch an emotion?

Or is it akin to an intuition, a memory of something that is about to happen, a knowing, a connection to the fabric?

Love and friendship bind humans together, which appears to help in countless ways, at least in this ecology and on this planet.

But in the dream world, perhaps, emotion is not needed.

Do you need emotion to read and understand an encyclopedia, Wikipedia, Google or Bing?

Does emotion disable the connection to "knowing" but enable connections in the "real" world?

Is this why there is emotion?

One hundred percent for Earth's community will not go away in the awake world.

It rings as a knowing, an uncluttered thought that just is, but is the connection true, the reception clear. That is always the question from the night.

It is persistent in the day. Emotion reigns, impossible to set aside.

It is what should be, one hundred percent for Earth's community.

What is good for Earth's community? Jobs, community, good housing, no poverty, no suffering, no hunger, opportunities, freedom, free will, quality goods at reasonable prices, no corruption, no crimes, sound law, sound law enforcement, safety and security, sustainability, a green planet, peace on Earth?

In a bad economy, is achievement of what is good for Earth's community compromised?

If so, then are organizations compromised, those of private enterprise, religion, academia and government?

No matter the pursuit?

In good times, a great economy, plentiful jobs and opportunity, are organizations enriched, whether of private enterprise, religion, academia or government?

One hundred percent for Earth's community.

Public policy 2.0. Earth policy 2.0.

Is it really this simple?

If, to a hammer everything looks like a nail, to a nail does everything look like a hammer?

In the night it is taunting in the day it is haunting.

A life as a shell.

Filled with nothing.

Not the good.

Not the true.

Not the beautiful.

Not the happy.

Questions linger, hundreds, perhaps thousands, perhaps one.

Plato expressed the pursuit of three ideals, the good, the true, the beautiful. The founding fathers added the pursuit of happiness which may be all three.

If the pursuit of individuals can be so defined can that of organizations as well?

Is the good the most good?

Is the true ethics?

Is the beautiful the body, mind, and soul, or is it nature, the landscape, the planet, the Universe?

Does the pursuit of ideals apply to organizations?

The dream meanders unorganized at least in a human sense, but somehow purposeful. Without interventions it might lead everywhere.

The National Mall in Washington DC is the nation's front yard. It has green grass, tidal basins off the Potomac River, reflecting pools, fountains, cherry trees from Japan, spectacular old fractally branched trees and National Monuments commemorating the Nations great people and history.

If there is one manmade space in the country that should be beautiful it is the National Mall. It is how the world sees the USA. It is where Americans see the USA.

It is a reflection of all who live there like a front yard is of all who live within the house sitting upon it.

The space of the National Mall is, indeed, beautiful, the monuments historic.

But it is deteriorating.

It is not the most beautiful or well-kept space in the country.

Four seasons resorts does beautiful about as well as any organization in the world.

The National Park Service maybe not be as good, maybe not as well funded.

The National Park Service is a great organization, it's just that Four Seasons has more experience creating beautiful, the National Park Service, more experience maintaining natural landscapes.

Where there is a need is there an expert organization?

Could there be a partnership?

It is airport security.

Prospective passengers put carry-on bags, slip- on shoes, belts, wallets, everything metal, everything liquid, and everything in pockets on a conveyor belt to be x-rayed.

On this occasion a passenger puts cash in a bin, plenty of cash, enough to wonder why any passenger carries so much cash. The passenger uses three other bins to place a laptop, an iPad, and liquids for plain examination.

All of the passenger's stuff makes it through the x-ray machines and the passenger begins the rather lengthy process of re-dressing belts, coats, and repacking electronics into carefully coordinated baggage.

Except on this occasion the passenger forgets one small item.

The cash.

The passenger probably does not normally carry cash.

The passenger leaves for the gate.

The passenger behind notices the cash and the back of its owner making way toward the airport gates for departure.

The passenger behind grabs all of his various and sundry belongings, then grabs the cash and runs toward the owner of the cash.

He hands the owner five hundred dollars in cash and there is a thank you.

Is this the true?

The night in sleep is a peculiar place, freed of the ecology and physics of the awake world.

Is it good, true, and beautiful?

Is it the opposite, sometimes both?

It seems vast, unending.

Everything if it isn't dreamed, if it hasn't been, could be.

Is this the Universe? The Universe of knowledge, the Universe of events, a collection of all events?

Einstein might not think so, but it could be.

When an organization is formed, does its goal that brought it into existence remain true?

Does its goal become to exist, to remain an organization, to continue, to survive and even thrive, no matter the goals that established it?

Sustainability first, goals second.

Is this politics?

Do politicians act for their constituency, for their party, for sustainability? In what proportion?

How about attorneys? Celebrities, CEOs?

Is anyone true?

Is any organization?

A few, more and more as cultures mature.

In the night there are no schedules. Time does not exist. Clarity of thought and enormous events happen in seconds or not at all. In the day the mind tries to organize thoughts around a schedule, around structured events, tries to make sense of them in the real world to determine what will work.

There is a call not of the wild, not of nature, not of the outdoors or anything of the outer world but of the inner world, the mind.

It hums from the moment of birth like a siren to the conscious.

It is indefinable but it is something of the true.

It seems to grow over a lifetime, over generations, over cultures.

It seems unstoppable.

Bill and Melinda Gates create a foundation.

Many of this ilk do, more and more.

Bono goes about resolving world poverty.

Warren Buffet donates 30 billion to the Gates Foundation. The Gates' and Warren Buffet team up to sweet talk many billionaires into donating half of their wealth to charities. It seems an easy sell, at least it should be.

Name the foundations, Ford, Rockefeller, Carnegie. Famous sports figures create foundations or make significant donations to charities. The list is long but not long enough.

Does the money no longer mean anything?

Did it ever?

In the Fall of a life what is there?

No longer the search for a soul, but a review of achievement.

What is the mark of a soul?

What was done?

What was not done?

What can still be done?

Contentment is exercise toward achievement, attaining achievement, and exercise toward more. In the Fall of a life there is contemplation of this, perhaps even in dreams.

Public policy 2.0 rises from the night, from sleep, from grogginess, from dreams. This is not an explanation of its origin just the best that can be understood.

It seems good.

It seems true.

Is it even beautiful?

Most of all it feels right, an answer to the siren.

It feels like it has happened, taken hold.

Crept in the still of night into the fabric of human culture.

Is this a knowing?

Or just a wish?

Whatever it is, it feels good.

Businesses exist to provide a public benefit, an axiom.

All organizations so exist, but this fact is sometimes lost in the effort to survive, in an effort to control, and many times is subsumed by the untrue.

The rational awake mind organizes a list, why?

- do the most good, optimize the benefit to Earth's community
- relinquish control to the Earth community benefit
- for any worthy pursuit there is always a related and greater Earth community benefit
- Earth community benefit is the organization's benefit
- a rising tide of public benefit will float the pursuit of all organizations
- in the pursuit of profit, pursue and create public benefit in equal or greater measure
- lobby and collaborate for the public benefit
- inspire competition for the public benefit

- suppression of competition devalues an organization
- for-profit organizations are well equipped to provide public benefit

Was this dreamed up? (inspiration)
Or just a dream? (meaningless)
Is public policy 2.0 the future, an ideal?
Or is it the past?
Can this never be, or...
Will it be?

Public Sourcing

T he internet.
*   Social networks.*
Each is just born?

Such a network already exists. It pervades the Universe except no one understands it, nor is there ready access. It is infinite and instantaneous. It seems it will not break down. It seems everything is there. Perhaps.

By turn, the internet is infant. Social networks are still just developing. They are clumsy, not focused and immature.

Participation in social networks is like watching cartoons for children; mildly entertaining, even appealing but unengaging, leaving a memory of a patch of time that can never be retrieved.

Social networks gather, seemingly, for no organized purpose other than the public introduction of individuals, a meeting place for bits and bytes. All the fuss about individual pronouncements, individual advertisements on a network that purports social attributes.

Networks of friends? Maybe. It appears, rather, a network of contacts to promote individuals in place of community. Social networks seem brilliantly organized to gain eyeballs.

The dream that creates connection, and might be the means of all interconnectedness, the means of all entanglement, casts an image of the child-like qualities of life on Earth reflected prominently in the cultures so far devised for the internet.

Is there a reason the web and internet culture grew around teenagers and young adults? In terms of subculture are these the low hanging fruit, the easy pickings? Moreover, it has been young adults that have led the charge in developing web sites and web-based businesses. Youth spent on technology.

Can this be true?

The awake mind creeps into dreams asking questions.

Asleep, are there any questions?

But the question is fair. There are web sites that provide useful content on human health, current events, emergency information, reference material such as dictionaries, encyclopedias, thesauruses, and of course useful sites that guide you to all information on the internet.

So the internet has become the most important and useful tool in human history.

And with this thought, which is an argument with the fabric, awake comes. The cocky, bold, falsely confident awake mind charges to a conclusion pushing aside the logic of the fabric making it fade to black, to dismemory.

Awake, in the light, the mind wins the argument with the dark, most of them, but should it?

This time a question lingers, as perhaps always there should.

Of course the internet is the most valuable tool in human history. But...

The internet is still in its infancy. For proof, review the vast changes that have occurred in the veracity of the companies that grew up on the internet, the interchange of them, the wholesale replacement of just about everything that was started initially. Most in just over a decade's time.

If it is in its infancy, there is vast room to mature. If it is in its infancy, it is juvenile, not at all what it will become, like all of life at the time of birth. (Does the internet obey the rules of nature, David Bohm's underlying implicate order?)

This is all the dream was saying. Violent agreement.

But there was something more from the dream. A calling.

The dream seemed to focus on the immaturity of social networks, specifically.

Awake, this is an interesting question.

What do they do?

What do they accomplish?

The better question, what can they accomplish?

Google, Craigslist, Facebook, Twitter (now X), Linkedin, Pinterest, Bing, Wikipedia, eBay and a host of others.

What can they accomplish?

What should they accomplish?

Is there a limit?

The dream continues, not in sequence, not the next day, not a week later, but sometime. Dreams obey no time, no sequence, more like a video, ney, a hologram of events out of any order, any sequence. (Do they mirror what the physical Universe is? Are they a portal to more than the physical Universe, more than three dimensions? Are they a portal to the real Universe, a window looking out from a holographic Universe?)

It is an amalgam of numerous dreams over time, maybe all of time. It is the incomplete mosaic, the idea not the implementation, the direction to move but not the guidance to where.

That is the internet.

That is the state of social networks.

Mirroring the disorganization of a free enterprise society dominated by human vices of ego, control, and untruths, the social networks have charged forward with content mostly to attract eyeballs and associated advertising revenue.

But isn't this backwards?

The internet is a fresh cloth stretching out in front of us, a clean sheet of paper. One with nearly unlimited capability.

What could be done with it?

What would you do if you could do anything, the dream calls out mildly trying to appeal to the senses of the awake mind?

That is the internet.

Millions of people are doing what they could do on the internet if they could do anything, well almost anything.

Companies can start in seconds and operate with nothing more than thought, or just about.

It is a wholesale revolution.

And yet little is new.

When television was invented and broadcasting companies formed there was no technology available to determine how much TV was being watched to base monthly bills on. So, it came down to advertising revenue.

Television was free except you had to watch commercials.

This was not evil since broadcasters offered free television in a fair exchange for ads. It was not evil because broadcasters had no other means of monetization.

With a choice, what did internet broadcasters do with the internet? What has been done in total with the greatest tool in human history, a blank canvass with room for any possibility?

What are the gateways to the internet? Are they few? What do they charge? Do you need a computer? A mobile phone? A subscription? With a multi-year contract? Do you need applications? What do they charge? Are apps necessary or about to be? What have you been manipulated to need if you want to watch TV, read a book, or listen to a song?

Is there control? By whom or by what? Why?

Is there truth?

Untruth?

Is Plato's the good, the beautiful, the true embroidered into the internet?

Is politics?

Terrorists?

Governments?

Acadmia?

Wal-Mart, GE, IBM ...?

Big media?

Little media?

It is all there, every bit of human culture, the internet used merely as a tool to churn what already exists.

It is a maddening scramble of ego and control, like a political campaign with money and nefarious intent behind every plot, every spin, every truth, every untruth. Is it manipulation for personal or organizational gain?

Is that all it is?

Yes, if that is what the human community already is.

It is a nightmare from which the awake mind disconnects. That is not what the internet is about. The awake mind, lucid in the dream, abruptly terminates it like so many other nightmares that rain shivers of sweat. Too much of that in the day, no need for it in the night.

The internet, the awake mind counters, is about innovation, new and improved ideas for products, new services, new resources, open architecture, shared ideas and access to information about anything. Mostly it is about access.

It will break down barriers, transmit across borders, equalize cultures, provide opportunity for all. Or, as the dream world implodes upon sleep, it will not.

It will not.

It will not.

The words repeat hanging in the early light of morning like a mist across a pond, obscuring the truth of the pond, its features, even its essence.

This is not a knowing the awake mind argues.

It is worse.

It has happened. It is not in the future, it is now.

Control has managed its infestation.

What if… the dream begins this time? What if things were different, if things could be made different on the internet?

What if the Earth's collective community, its collective cultures could be affected by the internet, tweaked here and there?

What could be done? As this notion flickers on and off over a series of unconnected nights, like a tortured neon sign at the end of its life, it becomes apparent that it is awake where resolution lies.

Somehow the nightmare, over time, is not. It turns into a dream remembered for possibilities, not what exists.

Something to get to work on. A rudder to change course.

A purpose.

That is the internet.

That is the connection of one dream to many, of many to one.

At night it is the siren call, a call to order, a call for the truth.

It can be ignored, even for a lifetime for those as Lee Standing Bear Moore described who hear but do not listen. But if listened to, it is difficult to ignore. With survival assured, it is as if these are 'execute' instructions. One hundred percent for the good of Earth's community.

What has been accomplished? What is there to accomplish?

Take the ideas, a clean sheet of paper, obey no boundary, stay to the true and innovate. Pick up the receiver, take the siren call from the voice in the night, from the fabric. Answer it.

In this awakeless time, deep in the night, there is no running, no where to hide. It is a story, a history, a collective history, a personal history. It is has happened, a knowing, but this night the story is discovered.

Is it the story that better we will do?

It exists in all of us somewhere and is played out over human history hundreds, even thousands of times. It is the story of the former PayPal founder, a daring entrepreneur who made millions helping to develop a secure internet payment system.

What to do with a pile of money? How about designing and building electric cars to save the world because Detroit has not done it. A startup from scratch to tackle the giant automobile conglomerates worldwide and move the world to electric and away from polluting energy sources.

One person, one pile of money. There are others.

There are dozens maybe hundreds even thousands of innovators. But out of tens of thousand, hundreds of thousand, millions, even billions of people it seems small, statistically meaningless and therefore an anomaly. When everyone is connected how can this be?

History has other examples of even greater feats. The founding fathers, Dr Martin Luther King, Ghandi, Mother Teresa...

But the names of even ten more are difficult, the anomalous few not commemorated as much as pop stars, sports stars, actors, politicians or even fictional characters. Do the media control the appeal to just a few? Do just a few control the media? Do specific ideologies control the media?

How many great deeds have been hidden, how many undiscovered? If statistically a negligent amount has been discovered, what remains undiscovered must include all of the great ones.

So much achievement lost in the folds of human history.

So the siren calls. In the night it is answered, awake is it merely wondered about?

It is an idea of ideas. It is an idea of art.

There is no plan or design among the world's organizations, its political leaders, or its various cultures to extract the best ideas or art.

It seems quite opposite. Ideas and art mean eyeballs and money but only if controlled absolutely.

Ideas and art are then imprisoned by the exercise of control of most organizations, political leaders, governments, academic institutions, religious institutions and the media.

In the night there is sweat, the dream, the inbetween times roiling with emotion, an anger of disachievement, a gaping hollow for what is lost or rather, what is not found.

Awake abruptly interrupts an evolving plan that was fleetingly evident before the dream turned into a nightmare.

The plan seemed simple, elegant. Take the randomness of the internet, let the wind brush it overnight and decipher the fractal that is produced as a result of the underlying rules of the implicate order of the Universe.

Awake, this is, of course, nonsense. There is no implicate order. Fractals are not an underlying rule and the internet not subject to it.

But the dream suggests that the internet is a fractal, or something like it. Fractals are an understanding, an order of randomness. Is the internet random?

Certainly, a billion good ideas are.

Awake, perhaps it is simple. If somehow a billion good ideas could be collected and order made of them, what might be achieved? If millions of minds could be focused on a single problem, and order made of them, what might be achieved? If millions of works of art could be discovered and order made of them, not by control but by grace, beauty, compassion and expression, what might be achieved?

Something grand is missing.

Ideas and art.

By the sheer number of undiscovered works of art, among the missing must be the greatest creative art works of all time. Never exposed, never distributed, never enjoyed and perhaps never motivated to be created. There is no design to extract these works of art, nor any plan to motivate the creation of impassioned ideas of the mind.

The world didn't conspire to create the least likely corporate, academic, religious, political and art's cultures for inspiring ideas, creative art and innovation from the world's collective citizenry, but it appears that is what happened. Business schools don't seem to try hard, if at all, to teach students how to cultivate ideas and innovation both within oneself and within an organization, or how business or organizational cultures inspire or impede ideas and innovation.

It is puzzling that most organizations of any ilk the world over seem more obstacle than facilitator for ideas and creativity. Is this the result of control, control by a few within the organization, and control by a few or one organization within an industry?

Few have pondered the scope of humanity's unlimited natural creativity collective worldwide cultures have laid to waste. The problem is not even acknowledged.

Within the fleeting remnants of the dream, or perhaps the interpretation of it from the inbetween times, there is an order, implicate or otherwise, of the randomness of ideas and art. A gathering of them is possible. A collection. A mirror on the fabric.

Can the internet gather and order them? Its child-like fanciful and juvenile qualities blur what is possible. So far it is mostly eye candy.

So what is it about the internet? What is so important?

It is human community and the planet's community. It is Earth's community? That is all there is in this tiny corner of the Universe. Perhaps that is all there is in any tiny corner of the Universe.

The internet is the next battle ground in the war for the good, the beautiful and the true. It is improvement opportunity, a chance to break down boundaries, take higher ethical ground than ego and control currently allow, determine absolute truth and make it available so that organizations of government, academia, business and religion can focus one hundred percent on Earth's community.

It is more than do no evil. It is a stage upon which humanity will decide how much good can be achieved.

Everything of the human psyche reduces to the need to provide value and from that value, achievement. Everyone must feel valued by deeds that create achievement. If not, the connection is missing, the fabric's siren nothing but whistlewind.

Everyone does achieve.

But there is not always value for the achiever.

Value, to be realized, festooned to the soul, must be recognized in measure equal to deed.

Value is almost purely social.

And what exactly is a juvenile internet good at?

If a book is written and nobody reads it, does it provide any value? It may have the greatest ideas in human history, or blueprints for great achievement but has it done anything if no one reads it? If no one reads it, is there no achievement? It is as if the ideas never were made, the blueprints never provided.

Does anything exist if it is not observed? Does anything exist for the collective if it is not observed by the collective? Perhaps all we need to do is observe (collectively) the good, the true, the beautiful and not observe anything else.

To value something, it must be discovered and valued by peers. It must have a utility. It must make things better.

Achievement is movement to a better existence.

But this is difficult awake. Ideas and art are not valued properly by most cultural, organizational or business models ever devised. Even if an idea makes its way to the top of an organization and gets implemented, and even if it makes the organization a billion dollars, what does the idea-maker get; a bonus and a promotion? While these might be important and prestigious, the organization's extreme riches in comparison devalue the idea and achievement. Acknowledgement and honor is almost never commensurate with achievement.

A glaring example is that of teachers, the purveyors of truth, the foundation upon which all ideas and art are founded. Teachers are paid very little for providing, arguably, the most value to the collective world citizenry. Moreover, their value is undermined by the constant budget battles and deficits inevitably related to the economy resulting in funding deficits for federal, state, and local governments during bad economic times. Little thought is given to the loss of a good experienced teacher to another higher paying career when a younger replacement teacher just out of college will cost half as much. Imagine the plight of the education the United States' and world's children receive when this happens all the time.

Is it imbued upon the collective cultures of the world to teach youth how to achieve for the benefit of Earth's community?

Has the internet provided a further venue to achieve value for idea-makers? The internet makes it easy to startup a business based on an idea. The value comes either in the successful struggle for monetization or, more likely, the purchase of the startup by a bigger organization (looking for control?) thus providing the idea-maker, and the world at large, with the true value of the idea. Of course, there is great risk in a startup, and most of the best idea-makers are risk averse so even the greatest ideas in the world likely never get off the blocks.

In the dream it is worse; a clearer knowledge of missing ideas.

Creative idea-makers are not usually risk-takers and risk-takers are not usually the most creative idea-makers. Consider that it is the risk-takers, the dominant organizational leaders that demand voice and action that normally move up the organizational ladder. By-and-large the risk-takers run the world.

They will awaken, if at all, later in life to the siren call, to deeds for Earth's Community and the greatest motivation for achievement. If wisdom takes hold, as it has for Warren Buffet and others, these individuals will run a profit or non-profit but hire idea-makers for vision.

The ideas and creativity of idea-makers, indeed the opus of their lives, are far superior to those of risk-takers, but alas, they are never destined to run the world. Ask any patent attorney about inventors. Their ideas and inventions are innovative, their hearts made of gold, but not a whit of business sense by-and-large. Inventions and ideas sit unused, unlooked at by those who could implement them while far lesser ideas are implemented within Earth's community.

The world is run, therefore, not by the best ideas or the brightest among us, but rather by the ideas of the risk-takers, flawed not only by the lack of creativity and genius but by lust for position, power and control mistakenly identified as achievement.

The result is more eye candy, not achievement.

Creative idea-makers and artists generally do not crave nor seek business success, fortune, fame, power or position. Achievement is pursued. A platform is needed to establish a culture that values creation, and a vehicle to achieve.

This does not exist yet. But can it?

It echoes, the dream world does, often times. The second and third times are haunting, diminishing yet true and even beautiful much like the echo above a bay on a lake in the Adirondack Mountains.

It is many dreams connected to one. It is one dream connected to many.

It is a true echo, resounding.

Does it resound to Google, Microsoft, Yahoo, Amazon, Apple, eBay, Facebook, X and other aspiring giants of the internet? Is there a blank sheet in front of them and are the hands of the idea-makers on the rudder?

Awake, the legs turn wobbly when the sum of these dreams begins to form into something understandable. It is a big idea made to fit in many tiny dreams that only seem to flicker like the distressed neon light. It is a puzzle-piece with Vanna White turning a giant letter every so often. With enough of them even the dimmest light bulb will glow... And then it hits like a locomotive and the weight of it is crushing, the dread now replaced with astonishment.

The burden can't possibly be carried. No one can do it.

But that is the point. No <u>one</u> can do it.

Can the internet tweak human culture; move it somewhere else, somewhere it has never been before?

If a single person can, if a single government or organization can, what can the internet do?

The answer from dreams has nothing to do with eye candy of a youthful internet. It is a matter of a platform for recognition of value and achievement.

Has the world been dominated by a top-down culture with antiquated technology, old-school management, old school thinking, where ideas and achievement could be devalued, bounded and even suppressed for preservation of position, power and control?

Ah, but the internet...

It is a fine instrument for public sourcing, as if designed to gather the orchestra and play a set of beautifully harmonizing symphonies.

Is the public no longer just a user and buyer of a product? Is this thinking antiquated, a relic of the dominant business personalities and organizations that grew up without the internet?

Should the public be involved in the ideas of a product, the creation of it, its marketing, its distribution, and of course, buying and using it?

The public already is... because of the internet.

But this function is still immature. It revolves around a central organization designing a product and selling it for a profit.

Why shouldn't it revolve around achievement, be based upon it?

Why is there such a fascination about Google? Because for a while Google exercised a different business model. It developed a product not for eyeballs or to make money, but a workable search product that would provide a great service to the public. The utility of it was Larry Page and Sergey Brin's vision. Later, such vision and the utility of search became clear, and the idea-makers were handsomely rewarded. Google did something valuable for the community of Earth.

So did Microsoft in its beginning. Bill Gates had great vision and understood before anyone else the value of software for productivity.

But has either Google or Microsoft continued that model? Has control replaced vision? You bet.

There is a blend of models buried in examples from the internet. It is one idea on top of another. The one on top cannot exist until the one on the bottom does. Neither might exist without the internet.

It all starts, though, with the primary components of achievement: (1) creating ideas and art which has two components; extracting already existing ones and motivating creation of new ones, (2) exposition of ideas and art so they can be, (3) found and recognized by those who, (4) value them and have resources sufficient to, (5) employ them which resolves issues and improves the Earth community which is, (6) the definition of achievement.

There is a purist model embedded in Wikipedia. Anyone can contribute an information article by creating it, editing it or adding to it. There is no advertisement, no marketing drive for eyeballs, no payment for contributors, but rather, a pure utility for the good of humankind. Its utility and public sourcing are sufficient sustenance. It is born of a vision that the public can create a better encyclopedia than any individual or organization.

Clever.

But the dream would not have it that way. Wikipedia does not match the components for achievement, or at least optimize them. Its brilliant concept is proof that public sourcing will produce the best product. But it does not measure nor compensate for the value of contributions. Over time can this devalues the ideas that make up the contributions? And control by the few has made its way insidiously into its content such that it no longer resembles a product produced randomly by the public for the best use of the public.

Like much of the traditional media, Wikipedia has ceded control to political ideology.

The dream, the inbetween times, do not run their course. Awake comes with a thought of Google. It comes with the thought of myriad ideas whose beginning can be attributed to the internet and whose end came with their purchase. The end provided a valuation of the idea, of the possible monetization of the idea.

Google, it is reported, understanding the new business model the internet created, offers vast monetary reward to its employees who dream up an idea that may produce a billion dollars in revenue. It does so to retain idea-makers.

It is another way to value ideas. The idea-makers are valued as much in this model as the company itself, for what is Google without ideas?

What is anyone, any organization, without ideas?

In most companies, the idea-maker gets a promotion, perhaps a bonus and a pat on the back for even the most incredible ideas.

These idea-makers now apply to Google.

And the idea-makers at Google are motivated because they are valued not just by financial reward, for they can see the wider benefit to Google and community that their idea achieves.

Google is the benefactor because it values ideas and the idea-makers who produce them.

But, if Google employs ideas a certain way, for the benefit of community in equal measure to company, everyone benefits, and the value achieved is true.

Wikipedia, though, has something that Google doesn't.

Public sourcing, although impeded by political bias.

It rings in the middle of the night echoing in the halls of the mind when the mind is emptied of the awake world.

Money motivates most of the people most of the time, but achievement motivates all of the people all of the time.

There is a business model somewhere between Google and Wikipedia. Perhaps it is merely the addition of public sourcing to the Google model.

Google does not collect outside ideas.

Wikipedia does not properly value the ideas it publicly sources.

It is so obvious asleep, but awake things don't fit so easily into a three-dimensional world.

Google would be public sourcing if it could, but somehow it cannot.

Wikipedia probably chose not to monetize its website for fear that advertisers might pay for broader listings and even faulty entries. Google did not have such convictions and remains, at least publicly, mostly unscathed by any claim of taking money for higher search placement. That said, it has been postulated by some that Google search has also been tainted with ideological and political bias.

The problem of public sourcing comes down to an ability to determine the best ideas or the best art.

There are many sites that publish novelists who have not been published through the traditional mass media. These sites are interesting ideas in and of themselves since they attract all sorts of books and ideas about books from otherwise unpublished and unread artists.

These sites, understandably, have millions of postings and there are now many of them. But the public has difficulty sifting through all of it because so much of it is not well done. Let everyone post and a lot of it will not be good. But limitations to posting, such as restricting posting to literary agents who sift for the good ones and get the same artists and material over and over, will not capture the best. The baby will be gone with the bathwater.

So, these sites are not well used by the general public, which in turn provides little income, and therefore little motivation for the unpublished author to post their good stuff on these sites. There is no value to create and post on these sites and control yet remains in the hands of the traditional publishing industry.

The internet has changed little about books, yet.

There are a few internet radio sites that allow users to create their own radio stations by selecting a favorite artist and having the site's software identify similar artists and similar music. This enables the listener to listen to music in a genre that they enjoy. The problem is that algorithms so far devised make mistakes and plenty of them in identifying similar music.

Even so, it is interesting and useful to be introduced to artists of similar ilk to one's favorites, and particularly some that are not even recorded by mainstream studios.

This is a step toward identifying the best no matter the source, as long as every artist is allowed to post their material to the site.

To get the best ideas and art, posting of all ideas and art must be allowed. This will mean thousands and even millions of ideas and art.

What is missing is an ability to recognize the best. This is the holy grail not only for giants of the internet but for idea-makers as well. Recommendation software found on many web sites that provide other possible purchases if you like a certain book or song is frequently not useful. Most recommendation software is limited to published artists and points to the most popular, most recent or even to books or songs that are discounted or need to be sold to reduce inventory. A feeble attempt, if any, is made to identify the best.

There was an article in the Washington post a decade or so ago about gaming software that a researcher developed in which two or more players were pitted against one another to determine what the other would indicate is beautiful among many pictures presented. It was free to play.

After millions of plays the database of choices that were determined to be beautiful could define beauty quite well from human sourcing. This could then be matched against any photo on the web and an algorithm devised to present the most beautiful of them.

It is not Google's model since it is not a pure algorithm and it is not Facebook's model that bases recommendations solely on humans (friends, acquaintances). It is a combination, an idea on top of an idea. Google's model needs humans to identify beauty and creativity and Facebook's model based on human recommendations is unable to consider everything.

It is interesting that the researcher's idea and game mentioned in the Washington Post article was further funded by two organizations. Google was one and the Department of Homeland Security the other. It is unknown exactly for what purpose Google made the investment, but the article guessed that DHS would use the idea to develop recognition software to quickly and thoroughly assist humans to identify dangerous items upon security scanning.

Would it be too soon to think about enabling posting of all ideas and all art, paying advertising revenue even for the posting, and then providing recommendation software for the best so the public can find what it likes?

So the public can determine what it wants rather than a few agents, a few recording studios, a few movie studios, a few directors, a few novelists, a few recording artists, a few art galleries...?

No longer would you have to guess at the best books of any genre. No longer would new novelists have to wade through endless submissions to be published. A few clicks leads to a list of books in ranking order of quality that match specific interests. So the good books are identified for you.

Is this better than the bookstore and better than traditional publishing? Maybe.

The value is provided by public sourcing that drives an algorithm to help identify the best, and then from the public's purchase of good art. It should provide mass markets for the best and most popular works of art, and niche markets across a long tail for works of art that have specific but less commercial appeal.

Isn't the public at large better able to determine both commercial and niche value since it is the public who will buy and read the book?

Shouldn't this concept draw out those with resources who can display great art or easily mass produce great art more readily and in greater numbers than traditional sources?

Wouldn't anyone be able to invest in new artists in the same way a studio or a publishing company might today, or in the same way investments are made in the stock market?

Would this expand reward and valuation of art?

And wouldn't something similar be achievable for ideas, ideas of anything?

Can this model work in all art forms? Can it work for business? Can it work for ideas? How about education, science, politics? How about everything?

Would this be a kind of eBay for intellectual property?

Would this be a kind of Facebook for ideas and innovation?

Would this be both?

Can this model for public sourcing be used to resolve problems? Why shouldn't politicians ask the public to come up with the best way to reduce crime, the best way to clean our streets, the best way to elect our leaders, the best way to debate and resolve political issues, the best foreign policy, and the best way to handle immigration and on and on?

Would the results be any different for world issues?

Could this unbound smokestacks and borders?

An 86 year old man in reasonable health is stricken suddenly with immobility. Doctors make the lazy and simple diagnosis that it is old age. When they finally discover it is the onset of Parkinson's and they make a particular drug available to the 86 year old, mobility and energy return immediately enabling a return to a normal self-sufficient life (because Parkinson's typically takes years to become degenerative).

If people posted medical records and case studies, and millions of such records were available to search through, and a sophisticated search algorithm could rank those most closely matching the conditions and symptoms of this 86 year old man, imagine that the correct diagnosis would be only a few clicks away. The message is all the more important because Doctors make diagnoses based on experience and case studies with patients and those reported in the medical media.

Imagine if all such diagnoses and case studies were available.

Imagine fewer incorrect, and faster correct diagnoses.

Imagine a significant improvement in the diagnoses and treatment of most illnesses.

Now imagine how everything else might evolve.

Of course, this is being done on many fronts on the internet today. No one wants to give up their ideas, their expertise, their ego in a world made dependent on individuals rather than community.

Perhaps Google is working on this. Perhaps not.

Perhaps it will be left to idea-makers to construct this grand lifescape for all of us to contribute to and prosper from.

The model is global and inclusive. Can it inspire the very best ideas and art of all time, and match them to those who could publish them or implement them enabling fast achievement and vast improvement of Earth's community?

It is simple, do the most good.

This idea of ideas is just one. There are millions and even billions more out there. Can they be discovered to move the world?

It is a message not of one dream but of many, not of many dreams but of one. The inbetween times melt into dark. Sleep, finally, is sound.

Fabric

It is the dream and has been from the dawn of humankind. One started by floating across a bay, a much smaller part of a lake and a very small part of the planet, of the solar system, of the galaxy, of the Universe.

In dreams, the inbetween times, even flashes and dreams awake, hints have been dropped in a constant thread embroidered into the fabric. Here and there, awake in dreams, asleep, the clues assimilate. If sown together they become something.

Everything, it seems, ties to it, especially thought. The threads are more similar than dissimilar.

Dreams cull information from the Universe so much more and true than what is commonly believed to be in dreams. Can this be so?

How is the information harvested? What is the connection the dream makes, what does it connect with and what exactly connects?

What is the source of infinite information from an infinite Universe?

Awake the logical deduction is that there is something to connect to. In our physical three-dimensional world there is not much else to base assumptions on.

There must be some sort of medium. If so, what is it and what is its purpose?

Information flow?

Energy flow?

Is it transmission; of particles... of bytes... of thought... of thought quanta... of biological quanta...?

Of consciousness?

Does this medium transmit instructions, or guidance, or some form of programming? Is it a receiver of thought and an (collective) observer of events, or even a data bank, an assimilator to chronicle reality and provide access to what is needed? Is it a kind of internet of the Universe?

Whatever its essence, it would appear to deliver or provide access to information on events impossible for human senses to otherwise receive.

Sometimes the information and events are relevant, sometimes not, some events have happened, and some have not, at least yet.

If everyone learns something from dreams that they could not know, not deduce, something they have never been exposed to before...

If everyone has such a "knowing" in dreams or daydreams, in the night, in the inbetween times...

Then a lot of information about a lot of events is accessible...

Perhaps it is all knowledge of all events, past or future?

Why would access to such a cache of knowledge of events be limited to some arbitrary subset of all knowledge of all events?

If all events of the Universe are accessible, past or future, does time exist at all? Is time only a construct reflected in our ecology simply as sequences of events, a means to structure an ecology, a civilization?

Does this lead to an entire theory of everything in the Universe?

A unified theory not from physics, not from laws of gravity, or motion, or relativistics, not from classical physics nor quantum theory, but from something else entirely?

Something missing from physics...

Something missing from all of science...

Something about the mind, about thought...

Something about the connection, the transmission and reception of information...

Something about the ether, the fabric that kneads across all corners of the Universe...

Something about the medium, the fabric, yet for science to discover...

Something about being two places at once, a physical world, a physical reality, and a thought world, the world of dreams, of interconnectedness, of instantaneous entanglement?

Is this consciousness?

Some sort of collective consciousness, the Universe's consciousness?

Could it be like programming from a cloud, where useful information about events exist, but also where instructions to execute specific functions, biological, physical or thought, exist and can be transmitted (downloaded)?

Is consciousness a force like the nuclear force, the electro-magnetic force or gravity?

Is thought?

Does consciousness exist in the medium, in the fabric, in the cloud?

Do humans have consciousness, or do they just connect to it like peripherals to a computer's central processing unit?

Are humans peripherals?

Whatever else the fabric may be, there is no real dispute about interconnectedness. This is understood asleep. The awake mind does not object. It acknowledges that the notion of interconnectedness has transcended all human cultures and all of human history.

But there is more than a connection. There is something of an intuition, of a knowing, of precognition, of instinct. Perhaps this is collective, a collection within the fabric, but put there by the receivers who may be transmitters as well. A collective pool of knowledge, insight and information, used and useful by the receivers and transmitters alike to help everything connected.

Everything transmits, everything receives transmissions and receptions, and the entanglement and influence of minds, of thought, of consciousness, of everything, for the benefit of everything, for the advancement of everything. Every small ecology, every large ecology in the Universe.

Give me an example demands a lucid intruding awake mind. There must be thousands even millions of examples.

Of course, how about the woman who saves her baby by moving the crib to another room the night a chandelier falls that would have killed the baby.

Only one example, how about millions?

How about when time slows down in an emergency and people do things in anticipation of events about to happen, seconds, for instance, before an automobile accident that is avoided.

No proof.

How about the far lower number of people on the 9/11 flights than usual for that time of year and that day of the week? This is statistically evident.

Still not millions.

How about déjà vu experienced by millions or knowing the phone will ring and knowing who is on the other end?

Still no proof.

How about one traffic fatality for every 100 <u>million</u> miles driven? A fact from the awake world. While one is too many, wouldn't statistics suggest this number is way too small? Consider who is driving after all; road rage, speeders, reckless drivers, aggressive drivers, older drivers, sixteen-year-olds, smokers, cell phone users, texters, those eating lunch, those reading, very uncoordinated drivers, folks with poor vision, and on and on. Consider conditions; nighttime, rain, snow, congestion, roads in ill repair, signs damaged or unreadable, deer in headlights and on and on. Given such circumstances, how might so few fatalities be explained?

No correlation.

But the Universe harbors no coincidences.

Still no correlation.

How about everyone in everyday life walking down a street knowing all kinds of things like how wide the sidewalk is without really looking at it? How about driving a car quite a distance without even thinking about it and wondering how you got so far without a thought about it? How about understanding what a movie will be before you see it, what a book will be before you read it, what a report contains before you read it?

How about those times when, almost subconsciously, you know what is in a report and so you do not read it to save time? There is a feeling of guilt but somehow the knowledge of the report is there.

Still not really evidence.

Take a report sometime that is not read, many times a subconscious non-action in most office environments. For proof, pick it up later. It says what was already understood. No need to spend time reading it. No need to feel guilty.

How can that be?

As Lee Standing Bear Moore might observe, it just is. It is not training, it is not learned, but it is knowledge. It is available to everyone except for the cloud of unbelievability that obscures vision of it. People grow out of any kind of belief of something fantastical and into the local ecology, bothered by busyness and a fabricated sequence of events necessary, it would seem, for biological creatures to endure. The drone of the fabric goes unlistened to. Its ever presence ignored.

In the dream there is no need of "local" evidence. What is there is true.

It is true for everyone.

Could the fabric be the carrier of the Universe's consciousness, perhaps even the (holographic) projector of the entire Universe and everything in it? Could it be a collective consciousness, the medium of connectivity, the medium of dreams...?

The fabric seems to fashion information of all events of the Universe, flow them instantaneously, entangle them from everywhere to everything.

It exists, the fabric, universally, like dark matter or gamma waves, or even energy quanta that may permeate all the four corners of the Universe. Is it an energy that kindles invisibly in the blank of space everywhere much like gravity?

In the dream there is a measure of truth, but awake these are only questions, a dreamer's contemplation.

Modern physics, the physics of quantum mechanics demonstrates entanglement – everything is tangled up with everything else.

Similar to the dream.

Similar to everyone's dreams?

Are dreams entangled?

More specifically, everything influences everything else; sometimes a large influence like the Sun heating up the Earth, sometimes a small influence, like the heat from a distant star.

The cornerstone of quantum mechanics that has been demonstrated in experiments is that reality is created, or at the very least directly affected by conscious observation. It has been difficult to accept that reality does not exist or cannot be directly affected unless there is observation of it from a conscious observer. This is completely counter-intuitive to the way we think.

Yet, quantum physics works. No prediction made by the theory has ever been found in error. It is a theory now understood to be basic to all of physics and thus to all of science.

Quantum physics demonstrates seeming magic. For instance, influence is instantaneous and events can dictate the past, essentially changing what already has occurred.

First, as soon as an observer observes something, that something becomes real and instantaneously influences or entangles with everything else in the Universe. So, it may become real to everything in the Universe at once.

Influence is instantaneous and does not obey the presumed laws of physics that nothing can travel faster than the speed of light. Apparently, awareness of this reality in a physical sense does not obey the speed-of-light limit (conscious awareness can be instantaneous).

Second, conscious observation of an event that makes it real immediately contributes to its history, or at least part of its history.

Does this suggest that there is no time element to the "real" Universe since what and how we observe in the present can change the past?

How does influence/entanglement travel instantaneously?

What is the medium of travel?

What is consciousness? What is it about consciousness that observers who must be conscious determine or greatly affect reality by observation?

These are profound questions that modern physics has no answer for.

Just as profoundly, why do seemingly random thoughts from dreams so parallel the discoveries of quantum theory? What is the interconnection?

Down through the millennia people have conjectured (dreamed?) about universal connectedness and now the concept is demonstrated.

What is the connection between dreams and the notion of entangled "influence?"

Do dreams travel in the same medium as quantum theory's "influence"?

Is it instantaneous information flow (transfer)?

Instantaneous interconnectedness?

Through dreams can all information be accessed?

Can dreams instantaneously access all observation?

Is this consciousness?

Awake the mind is scurried by the dark, the unfinished and unremembered dreams.

Yet a thought emerges from chaos.

The mind is a device born and raised on and therefore designed for planet Earth.

It learns from experience, sequential experience. It adapts to help its host survive on this planet, in this environment, in this ecology.

Is the mind a biological computer, one that can learn?

If a colony on Mars is established and scientific devices are placed there to view the planet, understand its weather, understand the nature of its red rock, its sands, discover whether it contains water and biological life, how would such devices be programmed?

Would these devices be programmed with all knowledge of the environment in which it will function?

Would it be programmed to learn from that environment?

For what purpose?

To function continually as programmed?

To continue functionality as long as possible?

To grow functionality?

To perform its function better and better?

To thrive, grow and improve?

Is this the mind?

In the middle of the night, as far away from dawn as dark, maybe...

But in the daylight there is life to live, often little time to ponder such externalities.

The mind helps with survival, it helps people do better, develop, progress, thrive and achieve.

It functions to improve the human condition in the ecology of its existence.

To so perform, does it need any knowledge of other environments, other ecologies, other worlds, other galaxies or even other Universes? Does it need non-local knowledge?

Does it need knowledge of future or past events, of far in the future or far in the past events?

Is the mind merely programmed to survive and thrive in the environment in which it will function?

If so, does the mind only have that function?

Is the mind merely a device programmed for a specific function like a printer programmed to print?

If so, is there intelligence or just programming, sophisticated programming that allows learning?

If so, does the program extend to all life or everything, animate or inanimate?

Does the program extend to the whole of the Universe?

If so, is there a connection of all things to one another, perhaps through some central processing device where one peripheral does printing, another does calculation and still another regulates the local environment?

Is the entire Universe so connected?

Perhaps in the manner of a wireless network?

Or like the internet?

Is the interconnection itself consciousness, or a stream of consciousness, or a connection to it?

Or merely an elaborate program or set of instructions that direct the course of events...?

Not the specifics of how to play the symphony but that a symphony be played?

There are no giants of the paranormal, no Galileos, no Newtons, no Einsteins at least so accredited or acknowledged.

So none visit in dreams.

Of its several elements that include Extrasensory Perception (ESP), Telekinesis and Precognition among others, all are demonstrated by the experience and "feelings" (knowings?) of those aware of such capabilities.

Known science is on the fringe of indisputably proving the existence of some paranormal phenomena. Traditional science cannot disprove their existence.

But disfavor, discount and disconsider, science can ably do and has ably done throughout history.

In the dream this night this seems a particularly frightening thought given the realm of missteps, misunderstandings, incorrect theories and complete alterations of the worldview of science that occurs every half century or so.

The mindset, the pattern of our thought, the programming for life on this planet at this juncture of events, the need for others to believe what we do, the need for acceptance all take hold. They do not let go.

Such notions dominate everywhere but perhaps no more so than in science where only the proven is true, only the observed, only events studied in precise experiments.

Night abates. The dream is going nowhere, the inbetween times merely unsettled, like a stomach the day after alcohol.

The paranormal is for movies, fantastical conjecture, and horror stories.

Even the name, paranormal, connotes a spooky feeling.

It is the opposite of science. It is not true.

But there is insistence in the night, a voice in the mind, a reception of consciousness that speaks of truth. Awake it is denied and ignored.

But truth denied roils, somewhere, in someone's conscious.

This time it is not a truth that roils, it is an incomplete truth.

Modern quantum theory.

Quantum mechanics works exceptionally well. Its practical, real world application ignites a good part of the world's economy since it is the linchpin of many electronic applications.

Yet, physicists do not quite know what is behind it. Students of physics are taught the practical application and they graduate and take jobs with tech companies, good, high paying jobs, where they ably employ it to make new gadgets that do incredible things, but no one knows how or why it works.

It is like using a computer program. You know how to use it, and what to use it for, but you don't know how or why it works. It is therefore like discovering the practical application before discovering the science behind it.

Physicists desperately want to know the truth behind it but are, at the moment, completely baffled. They do not know why a conscious observer and the act of conscious observation are required to determine (create?) reality. They do not know how instantaneous "influence" (entanglement) happens. Such a dilemma has caused some physicists to invite conjecture even from non-physicists on the meaning of consciousness and its role in quantum mechanics.

Physicists are loath to admit such shortcomings, but some speculate that a new worldview may be necessary and that such a worldview is likely to be astonishing.

Others are skeptical and find it difficult to believe a new worldview is needed. Einstein himself called quantum mechanics and particularly instantaneous "influence" "spooky actions."

Physicists Bruce Rosenblum and Fred Kuttner explain in their book "Quantum Enigma" that if paraphenomena were convincingly demonstrated, "we would know where to start looking for an explanation: Einstein's 'spooky actions.' Going a bit further, the existence of quantum phenomena expands the scale of what is conceivable and thus increases the *subjective* likelihood of paraphenomena. (We're using "subjective" in the Bayesian probability sense.) The very unlikelihood of paraphenomena within the present physical theory means that its confirmation, no matter how weak an effect, would force a radical change in our worldview."

According to these scientists, ESP is the acquisition of information by some means other than the normal senses, for example, mental telepathy or remote viewing. Precognition is being able to discern what will happen in the future. Psychokinesis (PK) is the causing of a physical effect by mental action alone, for example Uri Geller's supposed "spoon bending" or the reported influencing of mechanical systems or of radio active decay.

Precognition, researchers indicate, occurs mostly in dreams. Information comes in dreams. Is precognition information out of time?

Everyone dreams. Does everyone have precognition?

If so, are dreams, or at least some of them, paranormal, or, is paranormal actually normal or at least common?

Precognition is of the future. It is unobservable, at least as observation is currently thought to exist.

But if there really is no time...

Only events...?

Is precognition just information of events that have somehow been observed? Are dreams entanglements of events observed by someone conscious? Are then dreams an instantaneous transmission of conscious observation, or the result of instantaneous transmission of conscious observation?

At night, does the mind plug into a medium of instantaneous transmission of observation, of infinite realities, of all realities?

Perhaps this empowers the mind to a wider understanding than could exist locally, informed by a wide Universe of observation (and thus reality?) capable of influencing local survival, greater livability, greater knowledge, greater observation.

Is there circular logic here?

Conscious observations are made that are fed into a medium of instantaneous transmittal and influence (perhaps by dreams) that are received by the fabric (if the dream medium is not also the fabric). In turn this expands the capability of any receivers (possibly everything in the Universe) who in collective turn make further (greater?) observations that are instantaneously transmitted.

The fabric may then be viewed not only as a repository of all events but a library harboring both deposits and facilitating withdrawals, that is harboring both receptions and transmissions. All elements of the Universe deposit observations creating a collective reality and expanding the fabric's dimensions and transmittal capability thus improving the expanse of possible transmissions and influence that in turn expands reality (the Universe) and improves the benefits that accrue to receivers (all of us).

and so on.

Maybe.

So, after all, it is a mystery, a box within which a mystery has been held unfolded forever like the box with the magic in it unopened by J.J. Abrams.

The mystery is a journey, open the box and mystery is resolved, and the journey toward it is over but perhaps another begins.

Should the box be opened? Should the mystery remain?

Answers are not evident, as yet undiscovered.

Do they lay out there, all around us, every night, every dream, every daydream, every flash an underlying implicate order, an order from higher dimensions, dimensions that limited senses cannot tune?

Are there five senses when we dream?

Probably not.

Does the fabric contain the code for the implicate order, the underlying rules for the Universe. Is the fabric the rules? Or is it the vehicle that carries the rules?

Does it carry nothing but the Universe's interconnectedness, the stream of consciousness of every living thing, of every thing animate or not?

Does this entanglement create the rules?

Does it create the implicate order, even one multidimensional?

Does everything, then, that entangles create observation and then does this create reality originating from the fabric, guided by a collective consciousness <u>from</u> the fabric?

Is this what the dream of an omniscient view above a bay on a lake suggested?

Perhaps.

When the dream ends such a suggestion is nonsense. But the threads of many dreams cannot be summarily dismissed. The awake mind is too curious for that.

Yet spoken out loud, such dreams invite ridicule, jeopardize life and the pursuit of happiness because they fall beyond the realm of what is explicitly evident.

There is little notion of what might be implicit, what might be an implicate order.

A two-dimensional creature will view objects in a three-dimensional world as straight lines. Such a creature would be incapable of viewing most of what exists in a three-dimensional world. What does a three-dimensional creature see of a four or more-dimensional Universe?

Yet there is entrapment in limited beliefs, limited vantage points from a three-dimensional world even if far greater truths await discovery.

The awake mind wins and whatever guidance signals exist from the fabric are employed locally. It must be so because Earth's community is local.

Everything else, particularly the kneaded dreams, fall outside the lines of what is believed.

The dreams are sometimes beautiful, more than any awake vision.

The dreams are true, at least asleep, purely so. The clouds of untruths do not exist.

The purity of vision is spectacular bridging dimensional divides allowing unlimited sight, unlimited belief, unlimited thought.

There is achievement here, of a vast collection of events, of information of ideas, of art.

Waiting to be collected, waiting to be observed, waiting to be discovered, perhaps designed to be.

Note:

There is a secret about this book. It was not locally conceived.

Perhaps you knew.

It came from the fabric, from the collection, every thought, every word.

It was slept on, dreamed about, thought about unconsciously, thought about consciously.

Over time, a long time...

Maybe a lifetime.

It wrote itself much like Stephen King described.

It was warbled together from the fabric, the book itself merely a bunch of thoughts received from the Universe much like Thomas Edison described.

It was a dream, and from within a dream, a remembrance, a knowing, much like Lee Standing Bear Moore described.

It was discovery...

Of the fact (event) of this book...

Of the thoughts and questions within it...

Of its eventuality.

It would appear that the book is some form of "knowing."

In the night in the dark, in dreams and the time between sleep and awake, there is reception. Imagery is clear, ideas are clear, the sweep of the Universe revealed in bits, in bytes, ever so much more than the day. All events are there, the entire fabric, the entire collection, the Universe's collection.

Every thing in the Universe is likely both a receptor and receiver. This would seem the direction of physics, the book, then, a mere glimpse from a tiny receiver; of fact, of fiction, of just a few of the ideas and notions within the collection...

Why this collection, why these particular images, why these ideas among an entire Universe of knowledge, of events among an entire Universal entanglement?

It is, of course, unknown and perhaps, at least for now, unknowable.

Explore.

Discover.

Mine the ideas, sift for the good ones, filter the not so good.

Find the truth. The journey is not easy, but it is more than worth pursuit. Unfolded mysteries are the highlight of humanity.

In dreams or awake.

Ask questions, all of them.

The answers are difficult, hidden, enfolded in a vast Universe.

But search for them, please do.

The reader may very well be skeptical of the questions and notions with implications raised in the book. If so, the book has done its job, the Universe as well.

Skepticism is valuable, except when it is not.

What does the reader "feel?" Perhaps this is the best measure.

For the more traditional scientific minded, some of the effects of paranormal phenomena, particularly the acquisition of knowledge from means other than the normal senses, have been scientifically studied and demonstrated. Please refer to the Institute for Noetic Sciences (IONS) and in particular, Dean Radin's book "Entangled Minds" that discusses in some detail many of the studies.

In addition, science has recently documented what is called a 'feeling of knowing.' It is described in the writings of Jonah Lehrer including his recent book "Imagine: How Creativity Works." Jonah Lehrer has a degree in neuroscience from Columbia University and worked until recently as a science writer who describes current science advancements related to the brain. He was also, until recently, a science editor at "Wired" magazine and has written numerous articles on the subject of ideas and creativity for many publications including the Wall Street Journal. (For the purposes of full disclosure, Jonah Lehrer's book "Imagine" has been withdrawn from the market by its publisher for fabrication of some quotations by Bob Dylan in that book. His quotations of neuroscientists, the studies they made, and the science referenced in the book have not been the subject of dispute at this writing.)

The science goes like this. In the right hemisphere of the brain is located a section thought to be responsible for insights. Even research scientists describe insights as flashes of ideas or thoughts that pop into the head seemingly from nowhere. From experimental research, scientists believe these insights emanate from remote associations that the right hemisphere of the brain is capable of making when the prefrontal cortex of the brain located above the forehead is not active or consciously thinking about any particular matter. The prefrontal cortex is the part of the brain that directs day-to-day activities related to the physical world and daily life, the center of the so-called conscious mind. Alpha waves are more prevalent in the right hemisphere when the brain is in a relaxed state and not focused on anything in particular such as when taking a warm shower or just after waking up and still half asleep. Mr. Lehrer indicates that increased alpha wave activity is demonstrated in numerous experiments for subjects that successfully solve 'insight' puzzles. The use of regular good old logic from the prefrontal cortex proves, for the most part, unsuccessful in solving such puzzles. In fact, it needs to be turned off or nearly so in order for solutions to pop into the head from the right hemisphere. The solution emerges in a flash where increased alpha waves and a momentary burst of gamma waves are noticed in the right hemisphere of the brain.

Science does not know where the alpha or gamma waves arise from, nor do they know their purpose or how they aid in the creative process. It is a mystery.

The suspicion seems to be that these are linked to remote associations that the right hemisphere is making to produce the insight. This version of the science assumes the process is centric and that the brain is making these associations internally, seeing some sort of association or pattern of events or facts, perhaps from obscure observations the brain has stored away over many years, to arrive at an aha moment.

Could there be another version of science, one where this process is not centric to the brain? If there is a universal connection, does the presence of gamma waves at moments of insight suggest the possibility of reception?

Gamma waves exist throughout the Universe.

Thomas Edison and other idea-makers over many years describe their aha moments as receptions.

Scientists believe that many if not all flashes of insight leading to significant revelations in every human endeavor including the arts, have been the result of right hemisphere associations. Scientists do not know how these come about but document such revelations almost universally as coming 'out of the blue.'

The 'feeling of knowing,' scientists seem to conclude, rises after a sudden flash or insight. There is a strong 'feeling of knowing' that the insight is absolutely correct, or that a solution that the insight points to is attainable. This leads to an obsessive hunt to 'prove' the insight or implement the idea. There is then a transfer of the idea from the unconscious right hemisphere to the conscious prefrontal cortex of the brain for the purpose of translating the idea into real world utility. A person, for example, will adopt an idea as important and correct, engendering not only a sense of discovery but a sense of urgency to implement it for practical purposes such as inventing a useful product, or for artistic purposes such as composing a song.

The implementation process can be tedious but is driven by a 'knowing' that the insight is important, and a solution exists that can improve the human or planet condition.

Is this then, scientifically speaking, achievement?

Science has not associated the "feeling of knowing" with precognition, yet.

Mr. Lehrer, in his book "Imagine," notes that scientists have demonstrated that relaxed states of mind increase alpha waves which in turn increase the likelihood of insights. Mr. Lehrer points to research that suggests an ideal moment for insights is just after waking up. "The drowsy brain is unwound and disorganized, open to all sorts of unconventional ideas. The right hemisphere is also unusually active." According to researcher John Kounios, "We do some of our best thinking when we are half asleep."

Indeed, Mr. Lehrer quotes world class cellist Yo-Yo Ma, "I think the best way to perform is when your unconscious is fully available to you, but you are still a little conscious too. It's like when you are lying in bed in the early morning. I always have my best ideas then. And I think it's because I'm still half-asleep listening to what my unconscious is telling me. But at the same time I'm not in the midst of some crazy dream, because then it's just crazy. I guess it's a controlled kind of craziness. That's the ideal state for performance."

Mr. Lehrer goes on to discuss how Marcus Raichle has demonstrated that daydreaming increases ideas and creativity. When the brain is bored, people daydream. This is a relaxed state. The mental process coincides with increased brain activities in different parts of the brain at the same time including the right hemisphere. The result is an ability to notice new connections.

But are these connections internal or external?

"Sleeping is the height of genius," so said 18th century Danish philosopher Søren Kierkegaard. What he might be alluding to is the place in dreams where the conscious and unconscious minds coexist and is fertile ground for thought and ideas. Here too, in the state of dreaming, where the prefrontal cortex is inactive, lie increased alpha waves.

Mr. Lehrer summarizes the point: "While the precise function of dreams remains unclear, there's increasing scientific evidence that they enable our creativity, allowing us to make all sorts of surprising connections." He states further that according to Sara Mednick, a neuroscientist at the University of California Santa Barbara, "the reason dreams are such an important source of creativity is that, once the uptight prefrontal cortex turns itself off, we are exposed to a surfeit of surprising connections and strange ideas. Most of these new ideas will be useless, of course, just the surreal babble of the dreaming brain. But sometimes, if we're lucky, we'll find our answers in the middle of the night."

Eli Lilly is a large and successful drug company; big pharma. But in the late 1990s Alpheus Bingham realized that the company was squandering vast sums of money searching for the next best-selling drug. The company's strategy was every company's strategy; finding and hiring the best qualified scientists and scientific researchers with the most experience.

But it was not working. Large investments of time and research funding did not lead to new blockbuster drugs.

So, Mr. Bingham broke every rule in the research business. In 2001 he created a website called InnoCentive. He posted the company's difficult scientific problems on InnoCentive and offered monetary rewards to anyone who could solve them.

Without any expectation of success, Eli Lilly and Alpheus Bingham found, to their complete surprise, that the public had answers and astonishing creativity.

Mr. Bingham had, almost unwittingly, accessed the greatest pool of talent in human history – the public.

InnoCentive became the first widely used website to public-source the most difficult scientific problems. It did so by ignoring the golden rules of corporate secrecy and fear that competitors might learn not only what Eli Lilly was working on, but the possible solutions as well. All corporate intellectual property heretofore had been kept close to the vest, each company a vestibule unto itself.

Eventually, InnoCentive evolved to allow other companies including monster corporations like Proctor & Gamble and General Electric among many others to offer monetary rewards for ideas to solve their problems. The site now includes problems with reward money from hundreds of corporations and non-profits in eight scientific categories. The problems are as diverse as the companies.

And the solutions keep coming along with the reward money.

The idea-makers, those that solve the problems, and that now number in the hundreds of thousands, are a disparate collection of people from all walks of life and from everywhere on the globe.

They are the public.

Everyone has a great idea.

What is peculiar is that solutions largely come from people without experience or training in the subject of the problem.

As Jonah Lehrer notes in his recent book "Imagine," "All of us contain a vast reservoir of untapped creativity. The desire to make something beautiful, to express our luminous sensations, is not a rare drive confined to those with artistic training. That desire is present in everyone. We don't notice this need because we constantly suppress it, because the timid circuits of the prefrontal cortex keep us from risking self-expression."

What is the conclusion then? That none of us should be smoke-stacked, that all of us should rotate through completely separate walks of life? Scientists should become writers and writers scientists, artists should sit on corporate boards and corporate executives should run art galleries?

Perhaps government employees should work for Google, Pixar, Apple and 3M in exchange for Google, Pixar, Apple and 3M employees working for the government in some predefined rotation intervals.

Do we need a platform for motivating by reward the ideas and creativity that everyone has inside them?

Is this a dream?

There is a sentiment among neuroscience and creativity researchers for pockets of genius that history has evolved from time to time. There was Florence Italy in the 14th century spawning a renaissance of art and science that included the likes of Michelangelo and Leonardo Da Vinci among many others. Then there was the renaissance of literature in 16th century Elizabethan England that provided a fertile environment for perhaps the greatest writer of all time, William Shakespeare, to develop his trade among many other now recognized world class writers.

The dawn of the internet is currently producing tremendous innovation in all walks of business life including the genius of Bill Gates (Microsoft), Sergey Brin and Larry Page (Google), Steve Jobs (Apple), Mark Zuckerberg (Facebook), Elon Musk (PayPal, Tesla, Space X), Pierre Omidyar (eBay, Omidyar Network), Jack Dorsey, Evan Williams and Biz Stone (Twitter) and literally thousands of others. The center of this genius era seems to be Silicon Valley in California.

The mantra is emblematic: Innovate or die.

These genius eras seem to be largely historical accidents where shifted government policy, changed socioeconomic conditions and new technology, by no intent, brought like-minded innovators in close proximity. In these places, at these times, there was motivation to create and great ideas that were willingly, and sometimes unwillingly, shared. There was also a fair share of implementers that facilitated the development of ideas; the wealthy in Florence that commissioned great artists, the construction of great new theaters in Elizabethan London in a regime of slackened censorship, and the venture capitalists in Silicon Valley making billionaires of those with successful big ideas.

How about intentionally creating these pockets of genius?

In Mr. Lehrer's book there is a striking quote by neuroscience researcher Paul Romer: "We do not know what the next major idea about how to support ideas will be." That idea, or at least the platform would seem to be the internet.

Mr. Lehrer concludes in his book, "We need to innovate innovation," and, "We have to make it easy to become a genius."

His book further concludes that everyone has the capability of producing great ideas. Innocentive just about proves the point.

Imagine, from a dream, from untrained minds, from dreamers from all walks of life, from people who have never talked with one another or shared any ideas, come similar ideas.

Ideas supported by science.

Established science seems to corroborate a fair amount of what is dreamed in the night or thought about half asleep.

But what about the rest of it? Will it be corroborated, the thoughts and ideas from dreams, from "knowings," the speculations of science postulated from the night, from out of the blue?

Perhaps this is enough to think about...

For now...

The Theory of Everything Else

By
L K O'Neal

Published by
Winds

TO:
    thought, ideas and creativity
    and to:
    my daughters
    their mother
    their grandparents

Don't miss out!

Visit the website below and you can sign up to receive emails whenever L K O'Neal publishes a new book. There's no charge and no obligation.

https://books2read.com/r/B-A-JIAH-OXDTC

Also by L K O'Neal

Away Home
The Theory of Everything Else

About the Author

I turned to full time writing after a career in the Energy business in the United States. In addition to 'The Theory of Everything Else' I've written an emotional tale of love, the power of place and a small hometown titled 'Away Home' and the first book of a three book series about an apocolyptical biological seige designed to enable the replacement of the world's many failed governments with a single world government. Due out the middle of 2024.

Read more at https://lkeithoneal.com/.